Introduction to Wireless System Design

Introduction to Wireless System Design

From Circuits to Web-based Applications

Henry Lau
Lexiwave Technology, Inc.
Lexington, MA, United States

Ludy Liu
Lexiwave Technology, Inc.
Lexington, MA, United States

Keith Chan
Lexiwave Technology (Hong Kong) Limited
Kowloon, Hong Kong

WILEY

Published by John Wiley & Sons, Inc., Hoboken, New Jersey.
Published simultaneously in Canada.

For general information on our other products and services or for technical support, please contact our Customer Care Department within the United States at (800) 762-2974, outside the United States at (317) 572-3993 or fax (317) 572-4002.

Wiley also publishes its books in a variety of electronic formats. Some content that appears in print may not be available in electronic formats. For more information about Wiley products, visit our web site at www.wiley.com.

Library of Congress Cataloging-in-Publication Data applied for:

Hardback ISBN: 9781394172221

Cover design: Wiley
Cover image(s): © Ameen Alassar/Shutterstock

Set in 9.5/12.5pt STIXTwoText by Straive, Pondicherry, India

Contents

About the Authors

Henry Lau earned his MSc and MBA degrees from institutions in the United Kingdom and the United States, respectively. With over 28 years of experience in designing radio frequency (RF) systems, products, and Radio Frequency Integrated Circuit (RFICs) in both Hong Kong and the United States, he has a strong background in the field. He served as a Principal Engineer at Motorola and Conexant in the United States, where he developed RFICs for cellular phones and silicon tuner applications. Mr. Lau holds six issued patents and has one pending, all related to RF designs, and he has authored a book on antenna design. Currently, he is the founder of Lexiwave Technology, a company based in Hong Kong and the United States that specializes in wireless, Internet of Things (IoT), and artificial intelligence (AI) solutions. The company designs and markets RFICs, RF modules, antennas, radar, and wireless solutions integrated with AI technology. Additionally, he has taught various RF-related courses internationally and has chaired two symposiums and three seminars focused on wireless and IoT technologies in Hong Kong during 2016 and 2017.

Keith Chan is a seasoned professional with a diverse technical background spanning mechanical engineering, applied computing, and business administration. With extensive experience in developing consumer, electrical and electronic products, and IT solutions, he has also crafted numerous IT-related courses. As a skilled system architect and developer for IoT systems, Keith has expertise in mobile application software, web server software, and database design. He holds a Higher Diploma in Mechanical Engineering from the Hong Kong Polytechnic, a BSc in Applied Computing from the Open University of Hong Kong, and an MBA from Heriot-Watt University. Additionally, he has completed diplomas in Digital Visualization and Interactive Multimedia Design & Application and Computer Game Production Technologies from esteemed institutions.

Ludy Liu is a highly accomplished Senior Software Engineer and versatile Software Architect with a robust background in designing and developing world-class embedded software and firmware solutions. With extensive experience in architecting software for digital signal processors (DSPs) and real-time embedded systems, she has successfully deployed advanced technologies, including advanced driver assistance systems and network video recording security systems. She has a proven talent for managing the full software development life cycle and excels in isolating and resolving complex system integration challenges. With a Master of Science in Electrical Engineering from the National Technological University and a Bachelor of Science in Electrical Engineering from the University of Manitoba, her contributions to the field include a publication in the IEEE Midwest Symposium on Circuits and Systems, a patent for an innovative method in semiconductor sensor technology, and presentations at international technical conferences.

Preface

Drawing on over 60 years of combined, award-winning experience at the forefront of electronics and information technology, we bridge the critical gap between hardware and software. In today's competitive landscape, true innovation in wireless communication happens at the intersection of these two domains, and expertise in only one of them is no longer sufficient. This book offers a practical, hands-on guide that provides a clear framework for this integration, written for engineers, programmers, and managers. We guide you through the entire technology stack—from the foundational hardware of electronic circuits and devices to the architecture of complex systems, and from low-level embedded software to sophisticated server platforms and mobile applications. With an emphasis on practical applications over dense theory, this book is rich with real-world design examples and insightful discussions, providing the comprehensive, cross-disciplinary knowledge required to develop superior products. It is designed to equip you with the understanding needed to build the robust, efficient, and innovative wireless solutions of the future. For those wishing to explore specific topics in greater detail, each chapter provides curated bibliographic references for further study.

This book is organized into 4 sections and 12 chapters, covering Wireless Standards, Hardware Design, Software Design, and Putting into Practice.

The opening section, Wireless Standards, consists of Chapter 1, which introduces the fundamentals of wireless technology in the Internet of Things (IoT). The chapter provides an overview of network topologies, major wireless standards, and the regulatory landscape of compliance and certification. Understanding these core concepts is essential for appreciating the role of performance and reliability in developing compliant IoT systems.

The Hardware Design section begins with Chapter 2, which focuses on receiver design in communication systems, emphasizing circuit blocks and system architectures. It examines essential circuit components and discusses various system architectures and design considerations, such as heterodyne, direct conversion, and image-reject receivers. This chapter offers a thorough understanding of receiver design, laying the groundwork for addressing the demands of modern wireless systems.

As a complement to receiver design, Chapter 3 focuses on transmitter design, considering the system as a whole and its performance factors. It explores critical building blocks like the oscillator, modulation techniques, and upconverters. Power amplification, output filtering, and other considerations are also examined. Different transmitter architectures, including direct conversion and heterodyne, are discussed. The concept of transceiver architectures is introduced, along with various configurations. This provides readers with a comprehensive understanding of transmitter system design, serving as a foundation for advanced concepts and techniques in future designs.

Chapter 4 provides an overview of a new radio, the software-defined radio (SDR) and its advantages over conventional radio systems. SDR relies on digital signal processing algorithms, reducing the need for analog hardware and improving reliability and spectrum utilization. The underlying concepts and architecture of SDR are explored, including key building blocks such as antennas, radio frequency (RF) front ends, and analog-to-digital and digital-to-analog converters. The role of software-defined antennas and the importance of the RF front end in signal transmission and reception are discussed. Signal processing functions in SDR systems, including modulation, demodulation, and error correction coding, are examined.

As one of the key elements in wireless systems, Chapter 5 focuses on antennas for handheld wireless devices, highlighting their crucial role in ensuring efficient and reliable communication. It explores the fundamentals of antenna radiation, parameters affecting performance, and different types of antennas commonly used in handheld devices. Additionally, practical design considerations for overcoming size constraints in portable devices are addressed. The chapter provides valuable insights for optimizing antenna performance in handheld wireless devices.

Chapter 6 provides an overview of printed circuit board (PCB) design for wireless devices, emphasizing their role in performance and functionality. It discusses key RF PCB design considerations such as layer stack-up, component placement, grounding, power plane design, and decoupling strategies. The chapter addresses challenges like high-frequency signals, signal integrity, power distribution, and interference mitigation, highlighting the importance of proper grounding and noise reduction techniques. It also covers design strategies for various circuit types, including intermediate frequency, RF, baseband, audio, and power supply circuits. Overall, the chapter equips engineers with the essential knowledge needed for creating high-performance PCB layouts.

As an opening chapter of Software Design section, Chapter 7 highlights the vital role of embedded software design in modern electronic devices, covering applications from consumer products to industrial systems. It defines embedded systems as specialized platforms optimized for specific tasks with constraints on processing power, memory, and energy. The chapter outlines the design flow, starting with requirement analysis and system architecture, and examines architectural styles like event-driven and real-time systems. It also discusses state machines versus real-time operating systems and factors influencing microcontroller or system-on-chip selection. By the end, readers will understand essential principles for effective embedded software development.

Chapter 8 examines key aspects of embedded software development, including optimization, development tools, debugging techniques, and testing strategies. It highlights the importance of optimizing software for real-time performance within limited central processing unit resources, addressing factors such as memory management and code efficiency. The role of development tools in streamlining software creation and testing is also discussed, along with effective debugging methods for diagnosing issues. The chapter emphasizes the necessity of thorough testing to ensure functionality and system integrity, aiming to equip developers with the essential knowledge and tools to enhance performance and reliability in embedded systems.

Chapter 9 offers a comprehensive exploration of the foundational concepts and technologies driving modern web development. It examines the key components of website creation, including the crucial distinction between back-end and front-end development. In the back-end section, the chapter delves into web servers, databases, and MySQL security, equipping readers with the knowledge to build robust and secure database infrastructures. The focus then shifts to front-end design, where core technologies like HyperText Markup Language, Cascading Style Sheets, and JavaScript are explored, along with front-end frameworks and responsive web design principles. Finally, the chapter addresses the critical topic of network security, ensuring readers understand the measures needed to protect web applications and data. Practical coding examples in the IoT context are provided throughout.

As a closing chapter of Software Design section, Chapter 10 offers an overview of mobile application development, with a focus on IoT applications. It defines mobile applications as software for smartphones and tablets, highlighting their role in providing services similar to those on personal computers. With advancements in big data, mobility, and cloud services, mobile applications have become crucial in modern information technology. The chapter examines tools, frameworks, and development environments for various platforms, including Android, iOS, and cross-platform solutions, while also addressing the integration of mobile applications with external services to help readers create versatile IoT applications.

Putting into Practice section comprises Chapters 11 and 12, both of which provide invaluable insights into the application of concepts and techniques in IoT and wireless solution development.

Chapter 11 presents practical design examples to highlight the importance of concepts and techniques discussed in previous chapters related to the IoT and wireless solution development. It focuses on two smart city projects: the Smart Lamp Post (SLP) and the Smart Radio Frequency Identification System (SRFID). As urban areas continue to grow, these technologies enhance efficiency, sustainability, and economic growth by optimizing systems such as energy consumption and transportation. The SLP integrates lighting management with environmental monitoring, while the SRFID improves asset tracking. These case studies demonstrate the integration of hardware, firmware, artificial intelligence (AI), and software to create innovative urban solutions. Readers will gain insights to apply these concepts in their own projects, empowering them to develop effective IoT and wireless solutions for modern urban challenges.

As the concluding chapter of the book, Chapter 12 explores significant advancements in distributed computing, cloud technology, blockchain, and AI, all of which play crucial roles in the expanding IoT landscape. It investigates the concepts of distributed systems and their industrial applications, highlighting the pivotal role of cloud computing in empowering IoT initiatives. The chapter also discusses the advantages of blockchain for enhancing IoT security and autonomy, while introducing AI algorithms that prioritize explainability and trust. Additionally, it examines the potential of edge computing to improve IoT latency and efficiency. Readers will come away with a comprehensive understanding of the transformative technologies shaping the future of distributed, connected, and intelligent systems, and their profound impact on the evolving IoT ecosystem.

Acknowledgment

Henry Lau
My sincere thanks go to David Michelson of Artech House for introducing me to Aileen Storry of John Wiley & Sons, Inc., which initiated this book project. I am also deeply grateful to Robert Alongi of the IEEE Boston Chapter for the invaluable opportunity to host courses and seminars in Boston, helping to shape this material. The content of this book is built upon the expertise gained from my past companies, colleagues, and customers, and I thank them for their role in my professional growth. On a personal note, I want to express my deepest appreciation to my son and daughter, Harmon and Fiona. Their patience and understanding were essential, and their suggestions and review of the manuscript improved it immeasurably. This book would not have been possible without them.

Ludy Liu
This book would not have been possible without the invaluable experience gained throughout my career. I extend my sincere thanks to my many coworkers for their collaboration and to the companies that fostered my professional growth. I am especially grateful to my parents, my two children, and my sisters, Sin Yee Liu, Siu Mei Liu, and Siu Yan Liu, for their incredible support and kindness throughout this process. Finally, I thank my good friend and mentor, Zhilla Breakstone, for the guidance and friendship. I am deeply appreciative of everyone who made this work a reality.

Keith Chan
The genesis of this contribution to the book began with an invitation from colleague and coauthor Henry Lau in 2022 to write sections focusing on IT applications for his comprehensive work on wireless systems. Accepting this challenge led to more than three years of dedicated effort culminating in this published volume, which is a gratifying achievement. A project of this magnitude requires a strong network of support. Appreciation is extended for the persistence that was tested throughout this journey. Sincere gratitude goes to all those who provided assistance along the publishing path, especially to coauthors Henry and Ludy for their

collaboration. Finally, heartfelt thanks to all members of the Chan family; their support and understanding were indispensable and deeply appreciated.

The Authors
We would like to express our heartfelt appreciation to Professor Henry Chung and Professor Andy Chan from the City University of Hong Kong, along with the dedicated staff from the Electrical and Mechanical Services Department of Hong Kong, Lexiwave Technology, Inc., and Lexiwave Technology (Hong Kong) Limited, for their invaluable collaboration on several of the practical projects discussed in this book. We are particularly grateful to Professor Andy Chan for taking the time to review some of the chapters. With the dedicated efforts and generous support from the editorial staff—Aileen Storrey, Ajikumar, Akhil, and Vishnupriya Rangachari of John Wiley & Sons, Inc.—, the book was ultimately and successfully published.

Part I

Wireless Standards

1

Wireless Standards

1.1 Introduction

As the first chapter of this book, this chapter offers a comprehensive exploration of wireless technology and its role in Internet of Things (IoT) applications. The widespread acceptance and adoption of wireless technology have opened up new possibilities across various applications. To establish a solid understanding of the significance of wireless technology in the IoT landscape, the chapter begins by providing a thorough overview of different types of wireless network topologies. In addition to examining network topologies, the chapter investigates the realm of wireless standards that are applicable to IoT applications. Prominent wireless standards such as WiFi, Bluetooth, Zigbee, LoRaWAN, and cellular technologies like 4G/5G are discussed, shedding light on their respective strengths, limitations, and specific use cases within the IoT ecosystem.

Furthermore, the chapter examines the regulatory aspects associated with wireless technology. It explores the regulatory requirements that govern the operation and deployment of wireless devices and networks. This includes topics such as radio frequency (RF) emissions, as well as other legal and technical obligations that manufacturers and operators must adhere to in order to achieve compliance and mitigate potential risks. Certification and type approval processes are also emphasized in the chapter, as they play a significant role in demonstrating compliance with the necessary standards and specifications. It highlights the importance of adhering to industry-specific regulations and provides insights into the processes involved in obtaining certifications. Additionally, the chapter discusses the regulatory bodies responsible for overseeing compliance and enforcing regulations in the field of wireless technology.

By comprehending the fundamentals of wireless technology and understanding the role played by wireless standards, readers gain insights into the significance of performance, reliability, and compliance in wireless communication. This chapter serves as a gateway to explore the intricate world of wireless standards and their

Introduction to Wireless System Design: From Circuits to Web-based Applications, First Edition.
Henry Lau, Ludy Liu, and Keith Chan.

profound impact on the design, implementation, and evolution of wireless and IoT devices and systems.

1.2 Wireless Technology

Wireless communication relies on the transmission of information through electromagnetic waves, eliminating the need for cables or wiring. The earliest wireless systems utilized radiotelegraphy, a form of radio communication that employed Morse code or other coded signals. These transmissions took place in the early 20th century [1]. As technology advanced, new modulation schemes emerged, enabling the wireless transmission of data, voice, and music through the medium known as radio.

A significant milestone in the development of modern wireless networking was the launch of ALOHAnet in Hawaii during the early 1970s. This groundbreaking technology laid the foundation for subsequent advancements and played a pivotal role in the creation of Ethernet in the 1970s. Additionally, it contributed to the development of the widely adopted IEEE 802.11 wireless standards [2], which have found widespread use in consumer, enterprise, and industrial applications.

The growing acceptance and adoption of wireless technology have led to the emergence of new application possibilities in recent years. The vast market potential of wireless technology has prompted a need to identify different types of enabling technologies for IoT applications [3–6]. This chapter will provide an exploration of the available IoT technologies, examining their respective advantages and disadvantages when deployed in practical devices and systems. Furthermore, it will address key electrical parameters and regulatory requirements, enabling designers to optimize system performance.

When designing an IoT device or system, customers often have specific requirements, such as maximum transmission distance, large bandwidth for handling substantial data loads, prolonged operational lifetimes, and cost-effective implementations. However, meeting all of these requirements simultaneously can be challenging, as they may involve trade-offs and competing considerations. This chapter will explore the important factors and considerations that must be taken into account to achieve optimal performance based on specific system requirements.

1.2.1 Types of IoT Technologies

The wide range of applications offered by the IoT has led to a revolution in various sectors [7, 8]. These applications can be broadly categorized into consumer, commercial, industrial, and infrastructure domains, each presenting unique benefits and opportunities. The success of these applications is heavily dependent on wireless technologies, which can be further classified into short-range, medium-range, and long-range communication technologies [9–11]. An overview of the most commonly used and promising wireless technologies is provided in Table 1.1, which will be discussed in more detail in subsequent sections.

In recent years, consumer applications of IoT devices have gained significant traction. These applications encompass a wide range of areas, including smart homes, wearable devices, and personalized health monitoring systems. IoT devices integrated within smart homes allow for remote control and automation of various household appliances and systems, resulting in enhanced convenience, energy efficiency, and security. Wearable devices, such as fitness trackers and smartwatches, collect and analyze personal health data, empowering individuals to monitor their well-being and make informed lifestyle choices.

Table 1.1 Common wireless standards and their capabilities

Type	Standard	Connection	US Frequency	Range	Modulation
GPS	Global Positioning System (GPS), Global Navigation Satellite System (GLONASS), Galileo, BeiDou	Point-to-multipoint	1575.42 MHz	20000 km	Binary Phase Shift Keying
WiFi	IEEE 802.11x	WLAN	2.4, 3.6, 5.60 GHz	100 m	WEP, WPA, WPA2
Z-Wave	Z-Wave	Mesh	314, 433, 915 MHz	30 m	Triple DES
Bluetooth Smart (BLE)	Bluetooth SIG	WPAN	2400–2483.5 MHz	50 m	GFSK
Zigbee	IEEE 802.15.4	Mesh	2400–2483.5 MHz	30 m	OQPSK
THREAD	IEEE 802.15.4/ 6LoWPAN	Mesh	2400–2483.5 MHz	10 m	DBPSK
RFID	ISO/IEC 18000-63	Point-to-point	13.56, 915 MHz	3 m	PAM
5G	5G NR	Point-to-multipoint	600 MHz, 2.5, 3, 24, 28, 30 GHz	100 m to 3 km	OFDM, MIMO
HSDPA/ HSUPA	3G	UTRAN	850/1700/ 1900 MHz	27 km/ 10 km	USIM
LTE	4G	Point-to-multipoint	700–2600 MHz	28 km/ 10 km	OFDM, MIMO
ANT+	ANT+Alliance	WSN	2.4 GHz	50 m	FSK

The commercial sector has also embraced IoT devices, utilizing them in retail, logistics, and transportation. Retailers make use of IoT technologies for effective inventory management, smart shelves, and personalized customer experiences. IoT-enabled logistics and transportation systems improve supply chain efficiency, asset tracking, and fleet management, leading to optimized operations and cost savings. Moreover, industrial applications of IoT devices have brought about significant transformations in traditional manufacturing processes, introducing concepts like Industry 4.0. IoT-enabled industrial systems leverage real-time data collection, predictive maintenance, and process automation to enhance productivity, minimize downtime, and improve overall operational efficiency. Industrial IoT applications extend beyond manufacturing and have applications in areas such as smart agriculture, energy management, and environmental monitoring, enabling sustainable practices and resource optimization.

Furthermore, IoT infrastructure applications focus on the creation of intelligent and interconnected urban environments. Smart cities utilize IoT technologies for efficient traffic management, waste management, public safety, and utility monitoring. By integrating data from various sensors and devices, these systems improve the quality of life for citizens, enhance sustainability, and optimize city operations. By understanding and harnessing the potential of wireless technologies in IoT applications, organizations and individuals can unlock the full benefits and opportunities offered by the IoT revolution.

1.2.2 Type of Networks

There are four major network typologies for IoT systems: (a) Point-to-Point, (b) Star, (c) Mesh, and (d) Mixed [12]. The choices made depend on the number of members/device nodes participating in the network.

Figure 1.1 shows the Point-to-Point network which has only two devices. Device one would be the end device in a typical IoT system collecting the data about its environment, including temperature, humidity, machine operation and motions, etc. Device two can be another end device collecting environmental data and exchanging between Device one. The communication can be one-way or two-way. In a traditional wireless communication system, Device two would be a base unit in a thermostat system controlling the temperature of a premise based on the temperature measured by the sensor inside Device one. For an IoT system, Device two would be the server computer or the cloud to which the end device is sending data

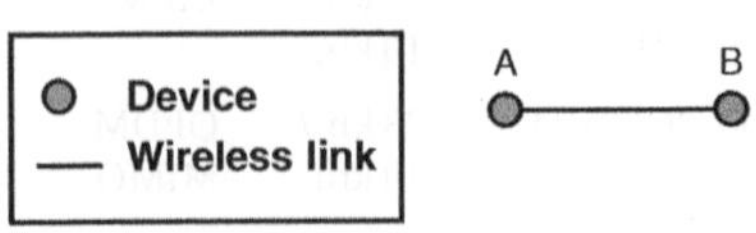

Figure 1.1 Point-to-point network

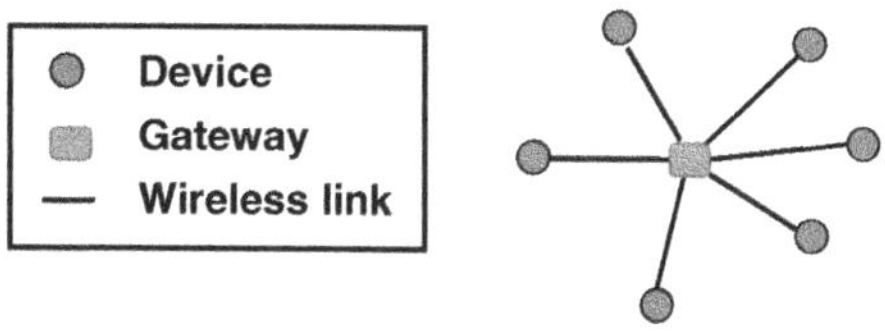

Figure 1.2 A Star network

through the Internet. The advantage of the Point-to-Point network topology is that it is very simple in implementation. The topology simply channels data either unidirectionally or bidirectionally between two devices.

The second network type is the Star network, as shown in Figure 1.2. A Star network has a gateway or hub in the center, which connects to all the end devices or nodes, while the end devices do not communicate with each other. The gateway transmits the data received from the nodes to the remote server computer or the cloud through the Internet. It also receives the data from the cloud and relays it to the corresponding end devices. Since each end device communicates directly with the gateway, the Star topology is also referred to as the "line-of-sight" architecture. The main advantage of Star technology is its simple design and potential low power consumption as all the complexity in the network is shifted to the gateway. All the end nodes need to communicate only in their allocated time or frequency slot. The actual communication could be time-division multiple access (TDMA), frequency-division multiple access (FDMA), or code-division multiple access (CDMA) [13]. The advantages of Star network are its simple, direct connections and ability to sleep between message transmissions to conserve power consumption, making it employed in many low-power IoT applications these days, i.e., WiFi and cellular networks. However, its main disadvantage is that the communication distance is limited, being between 30 and 100 m for low-power IoT applications as the transmission power in the radio will be low to extend the battery life.

The third network topology is the Mesh network as shown in Figure 1.3 where each end device communicates with each other. Messages hop from node to node until they reach the designated gateway with the capability of multihopping. Typical applications include home automation, smart buildings, and smart manufacturing using wireless standards of Mesh network topology such as Z-Wave [14], Zigbee [15], and 6LoWPAN [16]. The advantage of this network is its easy expansion of the system. There is no need to add new gateways as the network grows because a new device can join the network as long as it is within the range of any existing node. Another advantage is its high fault tolerance as each wireless node has multiple transmission paths between other nodes and the gateway. If one path is blocked or dropped, another path will be established automatically. Messages can still reach the gateway using other transmission

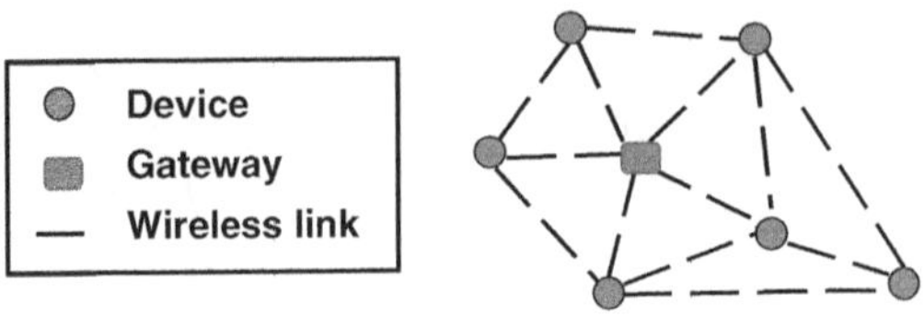

Figure 1.3 A Mesh network

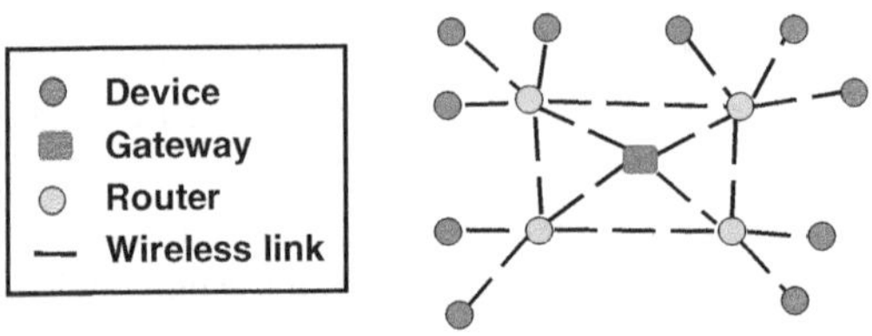

Figure 1.4 A Cluster-Tree network

paths. The disadvantage of a Mesh network is that the power consumption is potentially higher as the end devices have to listen continuously at a high duty cycle for messages from other end devices. The network latency will increase as the number of hops increases for a large Mesh network. The associated control software is more complex to create and maintain the routing table.

The fourth type of network is the Cluster topology, which is a combination of Star and Mesh typologies as shown in Figure 1.4. The end devices are connected in a Star network with the router while the routers and the gateway are connected as a mesh with each other. This Cluster network offers the advantages of both Star and Mesh networks of having low-power consumption on the Star portion and extended range and fault tolerance from the Mesh portions of the network. Despite all the advantages, the disadvantages of a Star topology remain.

1.2.3 Wireless Standards for IoT Technologies

There is a wide variety of wireless technologies being employed for IoT applications, including both proprietary and open-standard options. Among the open-standard technologies, some are widely recognized and commonly used in daily life. For example, Global Positioning System (GPS) [17], Bluetooth [18], cellular 4G/5G [19], and WiFi have become familiar to many people. However, there are also lesser-known open-standard technologies, such as Zigbee, LoRaWAN [20], Sigfox [21], and Narrowband Internet of Things (NB-IoT) [22], which are gaining traction in commercial and industrial applications. Each standard has its own set of advantages and disadvantages. Some offer larger transmission bandwidths, while others boast longer transmission distances. To help designers make informed decisions, Table 1.1 provides a summary of the key electrical and system parameters of these technologies. In this section, each of these standards will be briefly discussed, and their characteristics will be highlighted.

1.2.3.1 Global Positioning System (GPS)

The GPS is a satellite-based navigation system developed and operated by the United States, which provides geolocation and time information to users worldwide. It consists of a network of satellites in space, ground-based control stations, and GPS receivers commonly found in devices like smartphones, car navigation systems, and handheld GPS devices. GPS satellites orbit the Earth constantly, transmitting signals that allow GPS receivers to determine their precise location. By receiving signals from multiple satellites and performing trilateration calculations, GPS receivers can determine their latitude, longitude, altitude, and accurate time.

When it comes to transmitting power, the specific design and generation of GPS satellites affect their power output. Typically, GPS satellites have a transmit power ranging from 25 to 50 W. However, it is important to note that GPS signals weaken as they reach the Earth's surface. The strength of the signal received by GPS receivers can be influenced by factors such as atmospheric conditions, physical obstructions, and the quality and sensitivity of the receiver itself. Despite the relatively low transmit power, GPS signals can still be received and effectively utilized by GPS receivers in most situations.

GNSS, or Global Navigation Satellite System, is a term used to describe a collection of satellite-based navigation systems that provide global positioning and timing services. Alongside the widely recognized GPS, as just mentioned, there are other notable GNSS [17] systems such as GLONASS, Galileo (European Union), BeiDou (China), and NavIC (India). These systems operate by utilizing networks of satellites in space that transmit signals to ground-based receivers. By receiving signals from multiple satellites and performing calculations, GNSS receivers can determine precise location, velocity, and time information. GNSS technology has a wide range of applications across various fields. It is used in vehicle, ship, and aircraft navigation, as well as in outdoor recreational activities. Additionally, GNSS plays a crucial role in surveying, mapping, geocaching, and powering location-based services on smartphones. It has become an indispensable tool for navigation and positioning across different industries and in everyday life.

1.2.3.2 Bluetooth

Bluetooth technology is a wireless communication standard created to enable short-range data transmission between electronic devices such as smartphones, tablets, laptops, headphones, speakers, and IoT devices. It operates in the 2.4 GHz frequency range within the unlicensed Industrial, Scientific, and Medical (ISM) radio bands. The ISM band is known for being license-free, meaning that devices operating within these frequency ranges do not require individual licenses for use. However, there are regulations and restrictions in place to prevent interference with licensed communication services. Nonetheless, it is this lack of licensing requirements that makes the ISM band popular for various applications.

Bluetooth allows devices to establish wireless connections without cables or a centralized network infrastructure. It supports both point-to-point connections and connections between multiple devices, forming networks known as piconets. The primary purpose of Bluetooth is to facilitate data exchange and enable various functionalities, including transmitting audio, video, images, files, and control commands. Bluetooth profiles define specific functionalities and protocols for different purposes, such as Hands-Free Profile (HFP) for phone calls, Advanced Audio Distribution Profile (A2DP) for high-quality audio streaming, and File Transfer Profile (FTP) for file sharing. Bluetooth typically operates within a range of up to 10 m (Class 2), although there are variations with longer range capabilities. Devices with compatible profiles can automatically discover and establish connections when they are within range. This process, called pairing, involves authentication and the exchange of encryption keys for secure communication. To minimize interference, Bluetooth uses a technique called frequency hopping spread spectrum [13], rapidly switching between multiple frequencies.

Bluetooth has evolved over the years, introducing newer versions with improved features. The latest version is Bluetooth 5.3 which offers enhanced data transfer speeds, increased range, and improved power efficiency compared to earlier versions. Bluetooth Low Energy (BLE) is a power-efficient variant designed for low-power IoT devices with limited battery resources.

In terms of data rates, Bluetooth versions vary in their capabilities. Bluetooth 1.x and 2.x supported data rates up to 1 Mbps, while Bluetooth 3.0 + HS introduced Enhanced Data Rate technology, allowing a maximum data rate of 3 Mbps. Bluetooth 4.x maintained the 3 Mbps rate, and Bluetooth 5.x further enhanced data rates, offering speeds of up to 2 Mbps. BLE supports lower data rates, typically ranging from 1 kbps to 2 Mbps, depending on the specific version and configuration.

Bluetooth devices operate within specific power classes that determine their transmit power and range. These classes are commonly referred to as Class 1, Class 2, and Class 3. Class 1 devices have the highest transmit power, allowing them to achieve a range of up to 100 m or more. These devices typically have a maximum output power of around 20 dBm (100 mW). On the other hand, Class 2 devices, which are the most commonly used, have a range of up to 10 m. They generally have a maximum output power of around 4 dBm (2.5 mW). Class 2 devices are suitable for most everyday Bluetooth applications, such as connecting smartphones to headphones or speakers. Class 3 devices have the lowest transmit power and offer a range of up to 1 m. These devices typically have a maximum output power of around 0 dBm (1 mW). Class 3 devices are commonly used in applications where devices need to be in close proximity, such as wireless keyboards or mice.

Bluetooth technology finds applications in various domains, including wireless audio streaming, hands-free calling, wireless keyboards and mice, game controllers, smart home devices, healthcare devices, and other consumer electronics products. Its ease of use, wide compatibility, and versatility have made it a popular choice for short-range wireless communication.

1.2.3.3 Cellular 4G/5G

Cellular 4G/5G technology refers to the fourth and fifth generations of cellular network technology, providing wireless communication and data transmission capabilities to mobile devices like smartphones, tablets, and IoT devices.

4G, also known as Long-Term Evolution (LTE), is the fourth-generation cellular network that offers faster data speeds and improved network capacity compared to its predecessor, 3G. It enables high-speed Internet access, video streaming, and advanced mobile services. 4G networks typically operate on frequency bands below 6 GHz. 5G, the fifth-generation technology, represents a significant advancement in cellular networks. It provides even faster data speeds, lower latency, and increased network capacity compared to 4G. 5G networks can operate on a wider range of frequency bands, including sub-6 GHz frequencies and higher-frequency mmWave (millimeter wave) bands. The higher-frequency bands in 5G allow for greater data capacity and faster speeds, while the lower-frequency bands provide broader coverage.

The transmit power of 4G and 5G cell phones can vary depending on factors such as the specific device model, network configuration, and regulatory requirements in different regions. Generally, cell phone transmit power is limited to ensure efficient network operation, conserve battery life, and minimize interference. For 4G cell phones, the average transmit power typically ranges from approximately 0.1 W (100 mW) to 0.6 W (600 mW). The actual transmit power of a 4G cell phone can be dynamically adjusted based on factors like network conditions, signal strength, and proximity to the cell tower.

When considering 5G cell phones, the transmit power can differ based on whether the device operates in the sub-6 GHz frequency range or the mmWave frequency range. In the sub-6 GHz range, which provides wider coverage, the transmit power is comparable to that of 4G cell phones, typically ranging from approximately 0.1 W (100 mW) to 0.6 W (600 mW). In the higher-frequency mmWave range, which offers faster speeds but a shorter range, the transmit power is generally higher, varying from a few hundred milliwatts to a few watts. As a general principle, the higher the operating frequency, the shorter the communication range due to the phenomenon of free space path loss [23].

Cellular 4G/5G technology has revolutionized mobile communications, enabling faster Internet access, improved reliability, and supporting various advanced applications such as augmented reality (AR), virtual reality (VR), IoT connectivity, and smart city infrastructure.

1.2.3.4 Wireless Fidelity (WiFi)

WiFi, which is short for Wireless Fidelity, is a technology that enables wireless communication between devices using radio waves. It eliminates the need for physical wired connections, allowing devices like smartphones, laptops, tablets, and IoT devices to connect to the Internet and communicate with each other. WiFi technology operates based on the IEEE 802.11 standard, which defines the protocols and specifications for wireless local area networks (WLANs). There are various WiFi standards, including 802.11b, 802.11a, 802.11g, 802.11n, 802.11ac (WiFi 5), and 802.11ax (WiFi 6), each offering different data transfer rates, frequency bands, and features.

WiFi operates on specific frequency bands, usually 2.4 and 5 GHz. The 2.4 GHz band provides better coverage and obstacle penetration but has slower data transfer rates. The 5 GHz band offers faster speeds but has a shorter range and can be more affected by physical barriers. The transmit power of WiFi devices, such as routers and WiFi-enabled devices, is generally lower compared to cellular networks. WiFi routers typically operate at transmit power levels ranging from 50 to 200 mW, but these values can vary among device models and manufacturers. WiFi-enabled devices have even lower transmit power, usually ranging from a few milliwatts to tens of milliwatts. The actual transmit power can be dynamically adjusted based on factors like signal strength, distance from the router, and power-saving settings.

A typical WiFi network consists of a wireless router or access point and WiFi-enabled devices. The wireless router acts as a central hub, connecting to the Internet and wirelessly transmitting data to and from devices within its range. WiFi-enabled devices, such as smartphones and computers, have built-in WiFi adapters that allow them to connect to the router and access the Internet. To establish a WiFi connection, devices authenticate with the network using security protocols like WPA2 or WPA3, ensuring data confidentiality and integrity during transmission. WiFi networks often utilize a distributed network topology, deploying multiple access points strategically to cover larger areas. This approach improves coverage and capacity compared to relying on a single high-power transmitter. By distributing access points strategically, WiFi networks can provide reliable connectivity across a given area.

WiFi has become widely adopted in homes, offices, public spaces, and various industries, enabling Internet access, media streaming, file transfer, and seamless connectivity to network-enabled devices. The demand for high-speed and reliable wireless connectivity has driven the development of advanced WiFi standards with improved performance.

1.2.3.5 Long-Range Wide Area Network (LoRaWAN)

Long-Range Wide Area Network LoRaWAN utilizes a modulation scheme called LoRa (Long Range). LoRa is a proprietary wireless technology developed by Semtech Corporation. It is specifically designed for low-power, wide-area networks (LPWANs) and offers long-range communication with low-power consumption. LoRa modulation is based on a technique called Chirp Spread Spectrum (CSS) [24]. In CSS, the frequency of the transmitted signal gradually changes over time, creating a chirp-like waveform. This allows LoRa to achieve long-range communication while operating within a narrow bandwidth. It also has good resistance to interference and can penetrate through obstacles such as buildings and vegetation, making it suitable for urban and rural deployments.

LoRaWAN operates in the ISM bands, such as 868 MHz in Europe and 915 MHz in North America, and is now extended to 2.4 GHz. The range of LoRaWAN can extend up to several kilometers, depending on the environment and objects present. LoRaWAN devices typically operate at low transmit power levels to achieve long-range communication while maximizing battery life. The exact transmit power can vary between different device models and manufacturers, but it is generally in the range of 5–20 mW for most low-power LoRaWAN devices.

A LoRaWAN network is composed of several key components that work together to facilitate long-range, low-power communication. These components include end devices, gateways, network servers, application servers, and cloud or backend systems. End devices are the fundamental elements of a LoRaWAN network. They are equipped with LoRa transceivers, which allow them to transmit and receive data. These devices are typically designed to operate on low power and can be battery-operated, ensuring extended battery life. End devices can take various forms, such as sensors, meters, trackers, or other types of IoT devices.

Gateways serve as the crucial link between the end devices and the network server. Their primary function is to receive LoRaWAN transmissions from the end devices and forward the data to the network server for further processing. Gateways are responsible for receiving signals from multiple end devices within their coverage area, ensuring connectivity and data transfer. Network server acts as the central management component of the LoRaWAN network. It plays a vital role in managing and controlling the network's operation. The network server handles essential tasks such as device authentication, data encryption, and message routing between the end devices and applications. It ensures secure and reliable communication within the LoRaWAN network infrastructure.

The application server is responsible for processing the data received from the end devices. It receives the data from the network server and performs specific actions based on the application requirements. The application server serves as the interface between the LoRaWAN network and the end-user applications. It can analyze the data, trigger actions or notifications, and facilitate integration with other systems or services. Cloud or backend systems play an important role in the LoRaWAN network architecture. They receive data from the application server and provide additional processing, storage, analysis, and integration capabilities. These systems enable advanced data analytics, visualization, and integration with other systems or services. Cloud or backend systems are instrumental in extracting valuable insights from the data collected through the LoRaWAN network.

Due to its long-range capabilities and low power consumption, LoRaWAN is well-suited for applications that require long-range connectivity, low data rates, and long battery life. It finds applications in various sectors, including agriculture, environmental monitoring, asset tracking, smart metering, and smart city infrastructure.

1.2.3.6 Narrowband Internet of Things (NB-IoT)

NB-IoT, also known as Narrowband Internet of Things, is a wireless communication technology that also falls under the category of LPWAN technologies. It is specifically designed to meet the requirements of applications that demand long-range connectivity, low power consumption, and the ability to connect a vast number of devices. Similar to LoRaWAN, one of the primary advantages of NB-IoT is its exceptional coverage range, which surpasses other wireless technologies. It can provide extensive coverage even in remote or underground areas, making it suitable for applications such as smart city infrastructure, industrial monitoring, environmental monitoring, and agriculture.

NB-IoT operates within a licensed cellular spectrum and utilizes existing infrastructure provided by mobile network operators. It employs a narrowband radio interface and operates in a frequency range that can be deployed in a licensed spectrum and guard bands within LTE (4G) networks. This efficient utilization of existing cellular networks enables widespread deployment of NB-IoT technology. Users of NB-IoT typically need to subscribe to a cellular network provider that supports NB-IoT services. This subscription-based model grants users access to the network infrastructure and connectivity management services.

Power efficiency is a crucial consideration in the design of NB-IoT. It supports deep sleep modes, allowing devices to conserve energy when not actively transmitting data. This capability enables devices to operate on batteries for extended periods without frequent recharging or battery replacement. The power efficiency of NB-IoT makes it an ideal choice for IoT devices deployed in remote or inaccessible locations. The maximum output power of an NB-IoT device may vary depending on the implementation but is typically limited to around 23 dBm (200 mW) for single-tone uplink transmissions and 24 dBm (250 mW) for single-tone downlink transmissions. The maximum data rate for NB-IoT is approximately 250 kbps, making it suitable for low-bandwidth applications.

Overall, NB-IoT offers a robust, power-efficient, and secure connectivity solution for a wide range of IoT applications. Its long-range coverage, low power consumption, and integration with existing cellular networks make it an attractive choice for connecting a large number of IoT devices across various industries.

1.2.3.7 Sigfox

Sigfox is a wireless communication technology also specifically designed for LPWAN networks to cater to the connectivity needs of IoT devices. It operates in the ISM bands, utilizing ultra-narrowband technology for long-range communication. One of the standout features of Sigfox technology is its exceptionally low power consumption, allowing Sigfox devices to operate with minimal energy requirements. This translates to extended battery life and prolonged device operation without the need for frequent recharging or battery replacement. Consequently, Sigfox is an ideal choice for applications that prioritize power efficiency. Similar to LoRaWAN, Sigfox provides an extensive communication range, covering several kilometers in urban areas and reaching tens of kilometers in rural environments. This wide coverage makes it particularly suitable for applications that require connectivity over large geographic areas, such as smart agriculture or asset tracking.

Sigfox technology is cost-effective, aiming to provide affordable connectivity for IoT devices. The infrastructure required for Sigfox networks is designed to be economical, and the communication modules used in Sigfox devices are generally affordable compared to other wireless technologies. This affordability makes Sigfox a viable option for IoT deployments involving a large number of connected devices. The network architecture of Sigfox follows a simple Star topology, where IoT devices communicate directly with Sigfox base stations. These base stations collect and forward data to the Sigfox cloud, which acts as a central hub for data processing and routing to the appropriate applications or services. This streamlined architecture enables efficient and reliable communication between devices and the cloud infrastructure. Sigfox operates as a subscription-based service, requiring users to subscribe to a Sigfox network operator to access the network and its services. Subscriptions typically include data transmission and connectivity management for the devices. This subscription model ensures predictable costs and simplifies the management of IoT deployments.

In terms of data transmission, Sigfox utilizes a relatively low data rate optimized for intermittent transmission of small messages, such as sensor data, status updates, or alarms. Each message is limited to a maximum payload size of 12 bytes. This low data rate allows efficient utilization of network resources while contributing to the overall power efficiency of Sigfox devices. To achieve long-range communication with low power consumption, Sigfox employs a unique frequency-shift keying (FSK) modulation scheme known as Ultra Narrow Band

(UNB). It operates at a low data rate ranging from 100 bps to 2.1 kbps. This low data rate enables Sigfox to transmit only a small amount of data from IoT devices to the cloud. The maximum output power is limited to 14 dBm (25 mW) for uplink transmission and 27 dBm (500 mW) for downlink transmission in the 868 MHz frequency band. In the 902–928 MHz frequency band, the maximum output power for uplink transmission is limited to 20 dBm (100 mW), and for downlink transmission, it is limited to 24 dBm (250 mW). These power, frequency, and data rate characteristics make Sigfox particularly suitable for battery-powered devices, enabling efficient long-range communication.

1.2.3.8 Zigbee

Zigbee is a wireless communication technology specifically designed for low-power, low-data-rate applications. It operates in the 2.4 GHz frequency band and is based on the IEEE 802.15.4 standard. Zigbee devices are typically limited to a maximum transmit power of 1 mW or 0 dBm in the 2.4 GHz frequency band. Zigbee's primary objective is to provide reliable, cost-effective, and energy-efficient wireless communication for various industries, including home automation, industrial automation, healthcare, and smart energy.

One of Zigbee's notable advantages is its low power consumption, enabling devices to operate on battery power for extended periods. It supports relatively low data rates, making it suitable for applications involving intermittent and low-bandwidth data transmission, such as sensor networks and remote control devices. Similar to BLE, Zigbee typically has a communication range of up to 10–100 m, depending on environmental factors and obstacles. This short-range capability is well-suited for localized networks within homes or buildings, where devices need to communicate with each other in close proximity.

Zigbee supports various network topologies, including Star, Mesh, and Cluster Tree. In a Star topology, devices communicate directly with a central coordinator. Mesh topology allows devices to communicate with each other, creating a self-healing network where messages can be rerouted if a device fails. Cluster Tree topology combines elements of Star and Mesh topologies, offering greater flexibility in network design. To ensure secure communication, Zigbee incorporates robust security features. It employs encryption mechanisms to ensure data confidentiality and integrity. Additionally, it includes authentication and access control mechanisms to prevent unauthorized devices from joining the network.

Zigbee has gained popularity in applications such as smart homes, lighting control systems, HVAC (Heating, Ventilation, and Air Conditioning) systems, and industrial automation. Its low-power characteristics, Mesh networking capabilities, and support for a wide range of devices make Zigbee a reliable and versatile wireless protocol for IoT deployments.

1.3 Regulatory Requirements

The development and deployment of wireless products are heavily influenced by regulatory requirements due to several important reasons. One of the primary purposes of these requirements is to ensure the safety and compatibility of wireless devices. Regulatory bodies establish standards that encompass various aspects such as electromagnetic radiation, electrical safety, and product durability. By adhering to these requirements, manufacturers can guarantee that their wireless products do not pose any risks to human health or property. Regulatory requirements for wireless products can vary depending on the country or region where they are being marketed and sold. However, there are some common regulatory requirements that are typically enforced for wireless products.

One important regulatory requirement is certification and compliance. Wireless products often need to undergo certification processes to demonstrate compliance with regulatory standards. This involves testing the products to ensure they meet specific requirements related to RF emissions, safety, electromagnetic compatibility (EMC), and other relevant parameters. Certification may be required by authorized bodies, regulatory agencies, or telecommunication certification bodies before the products can be legally imported or sold.

Another regulatory requirement is frequency allocation and spectrum management. Regulatory bodies allocate specific frequency bands for different wireless communication systems to prevent interference and optimize spectrum usage. Wireless products must operate within these allocated frequency bands and adhere to power limits and other restrictions set by the regulatory authorities. Some products can be used in multiple frequencies and the manufacturers or developers should examine the requirements thoroughly so their product's ultimate operating frequencies have the optimal performance and cost.

Safety requirements are also a key consideration. Regulatory requirements focus on the safety of wireless products to protect users and prevent potential harm. These requirements address aspects such as electrical safety, product durability, and protection against electromagnetic radiation. Compliance with safety standards ensures that wireless products do not pose risks to human health or property. Failures in meeting the safety requirements will have huge consequences.

EMC requirements are another important aspect of regulatory compliance. EMC requirements aim to ensure that wireless products can operate without causing interference to other devices or services. Manufacturers must design their products to meet EMC standards, which include limits on unwanted electromagnetic emissions and susceptibility to external interference. Regulatory bodies also set limits on the amount of RF energy that wireless products can emit. These RF exposure limits are established to protect human health from excessive exposure to electromagnetic radiation. Manufacturers must ensure that their products comply with these limits and provide appropriate labeling or documentation to inform users about potential exposure.

Labeling and documentation obligations are often part of regulatory requirements for wireless products. This may involve providing information such as product specifications, compliance statements, safety warnings, and user instructions. Proper labeling and documentation ensure that consumers have access to essential information about the product's compliance, performance, and usage guidelines.

With the increasing connectivity and data exchange capabilities of wireless products, privacy and data protection regulations are becoming more important. Manufacturers may need to comply with requirements related to the collection, storage, and processing of user data, ensuring that appropriate security measures are in place to protect users' privacy. Regulatory requirements may also address accessibility and universal service aspects of wireless products. This ensures that they are inclusive and provide equal access to individuals with disabilities or special needs. These requirements may involve features or functionalities that support accessibility, such as text-to-speech, closed captioning, or compatibility with assistive technologies.

1.4 Certification and Type Approval

As one of the most important parts of the regulatory requirements, certification and type approval are essential for developing wireless products as they ensure legal compliance, verify product quality, promote interoperability, facilitate market access, build consumer confidence, and enhance safety. These processes contribute to the overall success and acceptance of wireless products in the marketplace.

Certification and type approval processes for wireless products involve the evaluation and verification of compliance with regulatory requirements. These processes ensure that wireless products meet the necessary standards for safety, EMC, RF emissions, and other relevant parameters. The important steps are listed as follows:

Certification Process:

a) Testing: Manufacturers typically conduct testing of their wireless products in accredited testing laboratories. The tests are performed to assess compliance with specific standards and requirements, such as EMC, electrical safety, RF emissions, and Specific Absorption Rate limits. The testing process may include radiated and conducted emissions testing, immunity testing, safety testing, and more.

b) Technical documentation: Manufacturers compile technical documentation that includes information about the product's design, components, test results, and compliance with relevant standards. The documentation also outlines the product's specifications, features, and user instructions.
c) Certification body: The manufacturer submits the technical documentation and test reports to a certification body or authorized third-party organization. The certification body reviews the documentation, validates the test reports, and ensures that the product meets the necessary requirements.
d) Certification issuance: If the product successfully meets all the requirements, the certification body issues the necessary certification or conformity mark. This certification demonstrates that the product has been tested and meets the prescribed standards. The certification mark is typically displayed on the product and its packaging to provide assurance to consumers and regulatory authorities.

Type Approval Process:

a) Application: Manufacturers submit an application for type approval to the regulatory authority or certification body responsible for approving wireless products in a specific jurisdiction. The application includes technical documentation, test reports, and other required information.
b) Evaluation: The regulatory authority or certification body evaluates the application and supporting documentation to ensure compliance with the applicable regulations and standards. They may review the test reports, technical specifications, and any additional information provided by the manufacturer.
c) Testing (if required): Depending on the jurisdiction and product category, the regulatory authority or certification body may require additional testing to be performed. This testing may be conducted by an accredited testing laboratory or by the regulatory authority itself. On the other hand, certain countries or regions may have established regulations that allow manufacturers to self-declare EMC compliance for their wireless products. This means that the manufacturer can perform the necessary EMC testing and evaluation internally and then declare that the product meets the applicable EMC standards and requirements. However, even in cases where self-declaration is allowed, there may still be requirements to maintain records of the testing and make them available upon request by regulatory authorities.
d) Approval decision: Based on the evaluation and testing results, the regulatory authority or certification body makes a decision on the type approval. If the product meets all the requirements, the type approval is granted, and the manufacturer receives the necessary documentation or certificate.

1.5 Regulatory Bodies

Regulatory bodies play a crucial role in ensuring compliance for wireless products. The specific regulatory bodies may vary depending on the country or region. For instance, in the United States, the Federal Communications Commission is responsible for regulating wireless communication devices and systems. They establish and enforce rules related to RF emissions, spectrum allocation, and equipment certification.

In Europe, the European Telecommunications Standards Institute (ETSI) is an independent organization that develops standards for information and communication technologies, including wireless products. ETSI standards cover various aspects such as interoperability, EMC, and safety. In Canada, Industry Canada is the regulatory body responsible for regulating wireless communication devices and systems. They establish rules and certification requirements related to RF emissions, spectrum management, and equipment compliance.

On an international level, the International Electrotechnical Commission sets global standards for electrical and electronic technologies, including wireless products. Their standards encompass safety, performance, and EMC.

Similarly, other countries have their respective regulatory bodies overseeing wireless products. For example, the National Broadcasting and Telecommunications Commission in Thailand regulates spectrum allocation, equipment certification, and compliance with technical standards. The Telecommunications Regulatory Authority in the United Arab Emirates is responsible for regulating telecommunications services and products, including wireless devices. In China, the Ministry of Industry and Information Technology oversees wireless communication devices and systems, setting regulations and standards related to spectrum allocation, equipment certification, and compliance.

It is essential for manufacturers and sellers to thoroughly research and comply with the specific regulatory requirements of the countries or regions where they plan to market and sell their wireless products. Adhering to these regulations ensures that products meet the necessary standards, are safe for consumers, and comply with the legal requirements of the target markets.

References

1 "What is wireless communications? Everything you need to know," Mobile Computing, last modified February 12, 2023, https://www.techtarget.com/searchmobilecomputing/definition/wireless

2 "IEEE 802.11," Wikipedia, last modified February 15, 2023, https://en.wikipedia.org/wiki/IEEE_802.11

3 "What is IoT?," Oracle, last modified February 17, 2023, https://www.oracle.com/hk/internet-of-things/what-is-iot/#iot-applications
4 Plant Web University. (2006) "Wireless Topologies," *Wireless* 105.
5 Links, C. et al. (2021). *Internet of Things for Dummies*, Qorvo 2nd Special Edition. John Wiley & Sons, Inc.
6 Vongsingthong, S. and Smanchat, S. (2014). "Internet of things: a review of applications & technologies," *Suranaree Journal of Science and Technology* 21 (4): 359–374.
7 "The enterprise Internet of things market," Business Insider, last modified February 25, 2015, https://www.businessinsider.com/the-enterprise-internet-of-things-market-2014-12
8 Perera, C., Liu, C.H., and Jayawardena, S. (2015). "The emerging Internet of things marketplace from an industrial perspective: a survey," *IEEE Transactions on Emerging Topics in Computing* 3 (4): 585–598.
9 Want, R., Schilit, B.N., and Jenson, S. (2015). "*Enabling the Internet of Things*," 28–35. IEEE Computer Society.
10 "The Internet of things: a jumbled mess or a jumbled mess?" The Register, last modified May 14, 2015, www.theregister.co.uk/2015/05/14/the_internet_of_things_a_jumbled_mess_or_a_jumbled_mess
11 "Can we talk? Internet of things vendors face a communications 'mess'," Computerworld, last modified April 18, 2014, https://www.networkworld.com/article/963923/what-is-iot-the-internet-of-things-explained.html
12 "Comparing mesh, star & point-to-point Topology in IoT Networking," last modified February 15, 2023, https://www.link-labs.com/blog/iot-topology
13 Rappaport, T.S. (2024). *Wireless Communications Principles and Practice*, 2e. Cambridge University Press.
14 Blokdyk, G. (2021). *Z Wave A Complete Guide*. 5STARCooks.
15 Elahi, A. and Gschwender, A. (2009). *ZigBee Wireless Sensor and Control Network*, 1e. Pearson.
16 Shelby, Z. and Bormann, C. (2011). *6LoWPAN: The Wireless Embedded Internet*, 1e. Wiley.
17 Kaplan, E.D. and Division, M. (2017). *Understanding GPS/GNSS: Principles and Applications*, 3e. Artech House.
18 Afaneh, M. (2018). *Intro to Bluetooth Low Energy: The Fastest Way to Learn BLE*, 1e. Novel Bits.
19 "3GPP (Third Generation Partnership Project) Specifications," 3GPP, last modified February 29, 2024, https://www.3gpp.org
20 "What is LoRa?," Semtech LoRa Technology, last modified February 29, 2024, https://www.semtech.com/lora
21 "What is Sigfox 0G technology," Sigfox, last modified February 29, 2024, https://www.sigfox.com/what-is-sigfox

22 Migabo, E.M., Djouani, K.D., and Kurien, A.M. (2020). "The narrowband Internet of things (NB-IoT) resources management performance state of art, challenges, and opportunities," *IEEE Access* 8.

23 Lau, H. (2019). *Practical Antenna Design for Wireless Products.* Artech House.

24 Springer, A. et al. (2000). "Spread spectrum communications using chirp signals," *IEEE/AFCEA EUROCOMM.*

Part II

Hardware Design

2

Receiver Design

2.1 Introduction

The design of a receiver is considered a critical aspect of any communication system. Receivers facilitate the capturing, extraction, and processing of information from received signals. In this chapter, the intricacies of receiver design will be explored, with a focus on the various circuit blocks and system architectures involved. An overview of the critical circuit blocks that constitute a receiver will be provided, including the antenna input and filter, low noise amplifier (LNA), mixer, intermediate frequency (IF) stages, IF amplifier, second IF circuitry, and IF detector. Each of these circuit blocks possesses its own unique function and requirements, which will be examined in detail.

Following the discussion on the circuit blocks, different system architectures and design considerations will be examined. The choice of system architecture greatly influences the overall performance of the receiver. The design considerations and challenges associated with various architectures will be explored. The heterodyne receiver, which utilizes the principle of frequency conversion to extract the desired signal, is one of the commonly used system architectures. The challenges of image frequency and image rejection in heterodyne receivers will be discussed, along with the concept of double-conversion receivers. Another approach to receiver design is the direct conversion receiver (DCR), which eliminates the need for IF stages. However, DCRs are accompanied by their own set of disadvantages, which will be examined. Furthermore, design examples will be provided to illustrate the implementation of DCRs. Lastly, image-reject receivers will be explored, with the aim of mitigating the challenges of image frequency. Various image-reject architectures and the use of complex polyphase filters in achieving image rejection will be discussed. Additionally, the disadvantages of low-IF image-reject receivers will be highlighted.

Introduction to Wireless System Design: From Circuits to Web-based Applications, First Edition.
Henry Lau, Ludy Liu, and Keith Chan.

By the end of this chapter, a comprehensive understanding of receiver design, including the critical circuit blocks, system architectures, and design considerations, will have been gained. This knowledge will serve as a solid foundation for the design of receivers that meet the requirements of modern wireless systems. Now, the exploration of receiver design will be initiated by examining the critical circuit blocks in the following section.

2.2 Critical Circuit Blocks

2.2.1 Antenna Input and Filter

To illustrate the critical parameters, a typical receiver is described first, block by block, to provide a basic understanding of the functions of all the individual building blocks. Figure 2.1 shows a conventional receiver, known as a double-conversion receiver [1–3], which has one of the highest performance levels among other topologies. Details about the double-conversion receiver, which is one type of the Heterodyne receiver architecture, will be discussed in more detail later in this chapter. The electromagnetic signal is picked up by the antenna and converted into an electrical signal. The electrical signal passes through the first radio frequency (RF) filter to the input of the RF amplifier. The first RF filter serves three purposes: (a) impedance matching [4, 5], (b) frequency selectivity, and (c) isolation.

Antennas are commonly designed with a reference impedance of 50 Ω, although 75 Ω is also frequently used in the very high frequency (VHF) to ultra high frequency (UHF) range. Depending on whether the antenna is used for transmitting or receiving, it is connected to the transmitter output or receiver input, respectively. For a receiver, achieving a conjugate match between the antenna impedance and the input impedance of the RF amplifier is necessary for maximum power transfer [4, 5]. However, due to various constraints or requirements, the terminal impedance may not always match the antenna impedance. In such cases, an

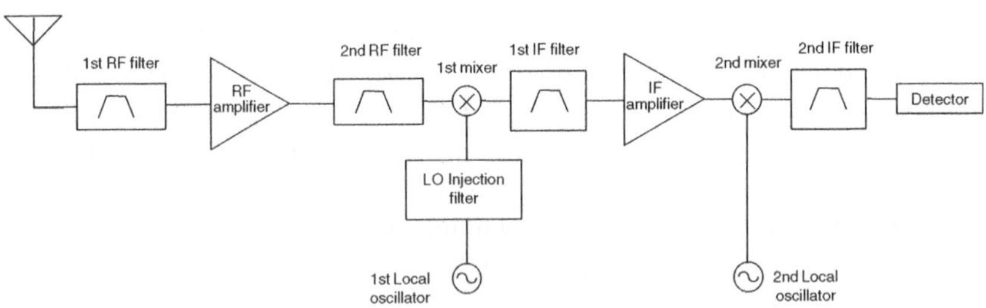

Figure 2.1 Double-conversion receiver

impedance matching network can be inserted between the terminal and the antenna to transform the antenna impedance to the conjugate of the terminal impedance or vice versa.

The impedance matching network is typically designed using either lumped or distributed elements, depending on the operating frequencies. In the case of lumped elements, reactive components such as capacitors and inductors are employed. Due to their frequency-dependent reactance, the impedance matching network also functions as a filter, providing frequency selectivity in addition to impedance matching. The frequency filter provides frequency selectivity, which is necessary to restrict the frequency bandwidth of the signal reaching the RF amplifier and the mixer. Without proper bandlimiting, large signals can overload the RF amplifier and/or the mixer, leading to intermodulation (IM) distortion [1] that impairs the desired signal quality.

2.2.2 Low Noise Amplifier

The output of the first RF filter is then passed to the input of an RF amplifier, also known as the LNA. Its main purpose is to amplify the incoming RF level to a higher level so that it can be detected and demodulated more easily into a useful signal at baseband for further processing. LNAs, being the initial active stage of receivers, have a pivotal role in the overall performance sensitivity, one of the most important parameters in receivers. It is crucial to note that the Friis receiving equation [6] provides insight into the sensitivity performance of a receiver system, which is a key aspect of its design. The sensitivity of a receiver is defined as the minimum signal level that the system can detect while maintaining an acceptable signal-to-noise ratio (SNR), as determined by the equation:

$$P_{\text{min}} = -174\,\text{dBm/Hz} + 10\,\log(\text{BW}) + \text{NF} + \text{SNR}_{\text{min}} \tag{2.1}$$

expressed in dBm

where

BW is the system noise bandwidth, and

NF is the noise figure of the system and is defined as:

$$\text{NF} = \frac{\text{SNR}_{\text{in}}}{\text{SNR}_{\text{out}}} \tag{2.2}$$

expressed in dB

where SNF_{out} and SNR_{in} is the signal-to-noise ratio at the output and input, respectively.

For a typical receiver as shown in Figure 2.1, it contains multiple stages of amplifiers, mixers, and filters. The Friis receiving equation governs the total noise figure in such a receiver.

$$NF_{eqv} = 1 + (NF_1 - 1) + \frac{(NF_2 - 1)}{AP_1} + \dots + \frac{(NF_n - 1)}{AP_1 \dots AP_n} \tag{2.3}$$

where

NF_{eqv} is the equivalent total noise figure,
NF_n is the noise figure of the n stage,
AP is the available gain of the n stage, and
all are expressed as ratios.

The dominance of the total noise figure can be attributed to both the noise figure and the gain of the initial stage in the receiver chain. Many resources on system design recommend prioritizing the optimization of the LNA as it typically serves as the first stage in the receiver chain. However, when considering a complete receiver, the antenna takes on the role of the initial stage in the receiver chain. In the case of a passive network like an antenna, the noise figure corresponds to the attenuation of the antenna. This attenuation can be regarded as the antenna's efficiency, which will be discussed further in the Antenna Chapter. Any slight decrease in the efficiency of the antenna will directly and significantly degrade receiver sensitivity. Therefore, system or product designers should give special attention to both the performance of the antenna and the LNA of the receiver.

The output of the LNA will go through the second RF filter to the input of the first mixer. The second RF filter serves the same function as the first RF filter to suppress receiver spurious response frequencies. The most trouble frequencies would be the IF, image (to be discussed) and second harmonic (due to the RF amplifier) frequencies.

The filter can also suppress the amount of local oscillator (LO) signal leaking to the antenna as shown in Figure 2.2. This phenomenon is typically called the LO leakage.

The LO signal is typically coupled to the antenna by the capacitances between the LO and RF ports of the mixer, as well as capacitances or resistances between the output and input of the LNA. Emitting LO power is undesirable, as it can

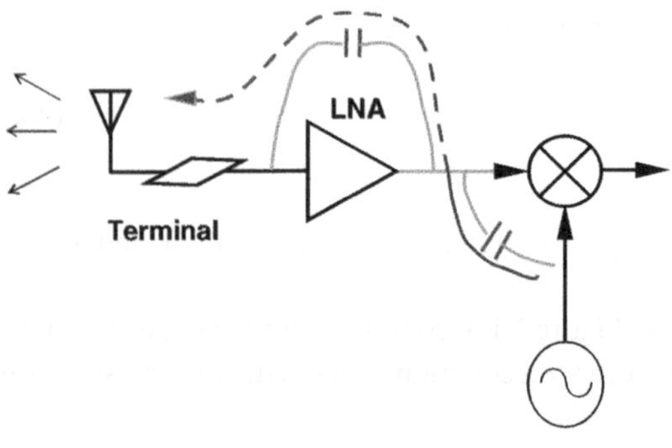

Figure 2.2 LO leakage

reduce the sensitivity of other receivers in vicinity operating in the same frequency band. It may also fail in the Electromagnetic Interference/Electromagnetic Compatibility (EMI/EMC) compliance requirement on unwanted emissions. Acceptable values for LO power emission typically range from −50 to −70 dBm, as measured at the antenna.

2.2.3 Mixer

Frequency bandwidth is often limited in many wireless communication systems across various countries due to factors such as a large number of users or the high cost of infrastructure required for building wireless systems. Consequently, it becomes necessary to restrict the channel bandwidth allocated to each user in order to fully utilize the available resources and maximize the number of users. For instance, the 3G GSM network only allocates a bandwidth of 200 KHz [7]. As indicated by Eq. (2.1), sensitivity is inversely proportional to the channel bandwidth. Narrower bandwidth results in higher sensitivity. Therefore, narrowband communication systems such as NB-IoT or LoRaWAN [8] can achieve higher sensitivity. However, due to the limited information rate defined by Shannon's theorem [9], sophisticated baseband processing techniques like coding, compression, and bandwidth-efficient modulation are essential for optimizing the utilization of available bandwidth.

The receiver must effectively handle the desired channel while simultaneously rejecting strong interfering signals both within and outside the desired frequency range. These interfering signals can be found one or two channels away from the desired signal, as depicted in Figure 2.3. Interferers may consist of unintended signals or useful signals intended for users on neighboring frequency channels.

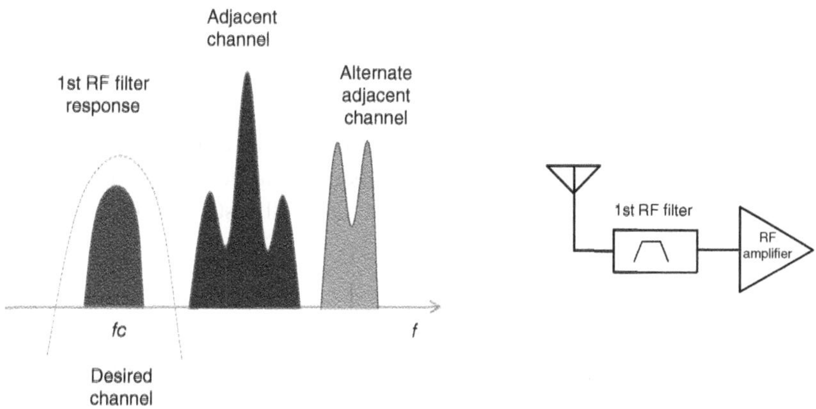

Figure 2.3 Channel filtering

However, it is essential for the filter to possess high selectivity (represented by a high Q factor) in order to reject signals from adjacent channels. Inadequate filtering can result in out-of-band interferers overloading subsequent stages of the receiver and causing a deterioration in signal quality. Furthermore, signal content near the edge of the adjacent channel can also impair the desired signal.

When the interfering signal is much stronger than the desired signal, with a difference of 50–60 dB, the filter's required quality factor [10] becomes unreasonably high. Additionally, the filter would need to have a variable center frequency to accommodate different carrier frequencies assigned to the user at different times. This poses challenges for precise implementation. Therefore, it is more practical to perform channel-selection filtering at a lower center frequency in the receiver chain, where the required filter Q values can be implemented.

While most receiver front ends include a "band-select" filter to reject interferers outside the desired frequency range (as shown in Figure 2.3), interferers within the desired frequency range are typically removed closer to the end of the receiver chain. This distinction between out-of-band and in-band interferers is crucial for suppressing components generated by users who do not comply with the required wireless standards and specifications. The passive band-select filter located at the front end of the receiver is faced with a trade-off between selectivity and in-band loss. Increasing the filter's order to enhance the sharpness of the band-pass frequency response requires incorporating more cascaded sections within the filter. However, as mentioned earlier, higher front-end loss directly contributes to an undesirable increase in the noise figure of the entire receiver. To mitigate this issue, the filter is designed to have a minimal loss of 0.5–1 dB while still providing some degree of frequency selectivity.

As previously discussed, performing channel-selection filtering at high carrier frequencies presents challenges. To overcome these challenges, a technique called "downconversion mixing" or "downconversion" can be applied. This method involves shifting the desired channel to a lower center frequency using a mixer. The mixer multiplies the radio frequency signal by a sinusoid $A_0\cos\omega_0 t$ generated by an LO, causing the desired channel to shift from ω_1 to ω_2, where $\omega_0 = \omega_1 - \omega_2$. The unwanted components at $2\omega_1 - \omega_2$ are then eliminated by a low-pass filter, allowing only the intended signal at the IF frequency ω_2 to remain. The choice of the IF frequency plays a critical role in the overall system performance, as trade-offs need to be carefully considered.

2.2.4 IF Stages

The output of the mixer is connected to the input of the IF filter. The initial IF filter, often made of crystal or ceramic, serves several functions. It protects subsequent stages from nearby IM signals, ensures selectivity between adjacent channels, and reduces the impact of the second image. The suppression of the second image is usually more critical than adjacent channel selectivity, and the number of poles in the filter is determined by the desired level of second-image selectivity. The equivalent noise bandwidth of the IF chain is important as it affects receiver performance by determining the amount of noise that reaches the detector and establishing the maximum modulation bandwidth that can be received.

In digital communication, maintaining low group delay in the IF filters is crucial. To mitigate the negative effects of group delay distortion, hardware- or software-based techniques can be employed for group delay compensation, as long as the group delay remains consistent across different units. The selection of the initial IF crystal filter after the mixer is important for achieving good IM performance and may require an impedance transformation network. This network limits the maximum impedance presented to the mixer on the filter's skirts. High impedance at the mixer's IF port can degrade its IM performance, especially when using active mixers.

2.2.5 IF Amplifier

The IF signal, after passing through the IF filter, is directed to an IF amplifier for amplification. The IF amplifier plays a vital role in the signal processing chain as it increases the strength of the modulated signal while rejecting unwanted noise and interference. Operating at a fixed IF frequency, which is typically lower than the frequency of the received signal but higher than the audio or baseband frequency range, the IF amplifier facilitates standardized circuitry for signal amplification and processing. The exact value of the first IF frequency in a double-conversion receiver can vary depending on the receiver's design and requirements. However, commonly used IF frequencies in double-conversion receivers are 10.7 and 21.4 MHz.

By converting the received signal to a lower IF frequency, it becomes easier to amplify and process the signal using standardized circuits. The IF amplifier amplifies this IF signal before it proceeds to further processing, such as demodulation or decoding. To achieve the necessary gain, the IF amplifier may consist of multiple stages of amplification, each providing gain to the signal. An important feature of IF amplifiers is gain control, which allows for adjustment of the amplification level based on the signal strength. Automatic Gain Control (AGC) is a common feature that adjusts the gain according to the amplitude of the input signal. AGC helps prevent overloading of subsequent stages and maintains a consistent output level despite variations in the input signal strength. Some IF amplifiers incorporate tuned circuits, such as inductor-capacitor (LC) circuits, to enhance selectivity and overall performance. These circuits resonate at the desired IF, allowing signals at that frequency to pass while attenuating others.

2.2.6 Second IF Circuitry

In double-conversion receivers (which will be discussed further in this chapter), the second mixer and second IF filter are added after the first IF circuitry. The second mixer brings the IF frequency to a lower value for channel filtering by the second IF filter. In some double-conversion receivers, the second IF frequency may be predetermined and fixed, such as 455 kHz or 10.7 MHz, similar to the IF frequency used in single-conversion receivers. While the design requirements for the second IF circuitry are generally less stringent, there are two important exceptions. First, the SNR at the detector input is crucial for specific receiver baseband performance. This detector property is referred to by various names, including co-channel rejection, capture ratio, and rise number. It is a critical parameter for any receiver design and should be kept as low as possible. Essentially, it represents the IF SNR at the detector input that is necessary to meet certain baseband specifications [6, 11]. Second, there is a minimum signal level that must be exceeded in order to extract the baseband signal through the IF detector. This minimum signal level is required for the proper processing and extraction of the baseband signal.

2.2.7 IF Detector

An IF detector plays a crucial role in extracting the original modulating signal from the IF carrier. Its primary function is to demodulate the IF carrier signal and recover the original audio or data signal that was modulated onto it. The specific circuitry of the IF detector varies depending on the modulation scheme used in the communication system. One common type is the envelope detector, which is used for amplitude modulation [12, 13] signals. It consists of a diode and a capacitor that rectify and filter the signal, extracting the envelope variations carrying the modulating information.

For single-sideband modulation [12, 13] signals, a balanced modulator or product detector is often employed. It utilizes a balanced mixer to multiply the IF signal with an LO signal, resulting in sum and difference frequencies. The demodulated audio or data is contained in the difference frequency, while the unwanted sideband and carrier signals are filtered out. In the case of frequency modulation (FM) or phase modulation (PM) [12, 13] signals, a phase detector is commonly used as the IF detector. This type of detector compares the phase of the IF carrier with a reference signal, such as a locally generated carrier. The output of the phase detector is a direct current (DC) voltage proportional to the phase difference between the two signals, representing the original modulating signal.

Following the IF detector circuit, additional stages such as audio amplifiers or data decoders are typically utilized to further process and enhance the recovered signal. These stages serve to amplify, filter, or decode the demodulated signal based on specific application requirements. In some cases, IF detectors may incorporate a carrier recovery mechanism, particularly for FM or PM signals that require accurate carrier synchronization. Carrier recovery circuits, such as phase-locked loops (PLLs), are employed to track and reconstruct the carrier signal for precise demodulation.

The accuracy of the demodulation process is essential in IF detectors as any distortion or errors introduced during demodulation can impact the quality of the recovered signal. Therefore, IF detectors are designed to minimize distortion and maintain fidelity to ensure accurate reproduction of the original modulating signal. IF detectors also influence the signal-to-noise ratio (SNR) of the demodulated signal. Well-designed IF detectors aim to maximize the SNR by minimizing noise contributions and optimizing the demodulation process. This is particularly important in low signal strength or noisy environments where the quality of the received signal may be compromised.

While the aforementioned types of IF detectors are common, there can be variations and hybrid approaches based on specific communication system requirements. For complex demodulation techniques like quadrature demodulation, which are used for demodulating quadrature amplitude modulation or phase-shift keying (PSK) [12, 13] signals, different approaches may be employed.

2.3 System Architectures and Design Considerations

2.3.1 Heterodyne Receivers

As mentioned previously, a heterodyne receiver is a type of radio receiver that utilizes heterodyning to convert incoming RF signals into a lower IF frequency for further processing. Heterodyning involves combining two signals to create a new signal with a different frequency. In a heterodyne receiver, the incoming RF signal is mixed with an LO signal, resulting in a heterodyne or beat frequency that represents the difference between the frequencies of the two input signals. The heterodyne frequency is typically chosen to facilitate the design and implementation of high-quality filters and amplifiers.

The main advantage of using a heterodyne receiver is that it simplifies the filtering and amplification of the RF signal. By converting the RF signal to a fixed IF, it becomes easier to design and implement high-performance filters and amplifiers that are optimized for that specific frequency. This leads to improved selectivity and sensitivity in the receiver. Heterodyne receivers are widely used in various applications, including radio communication, radar systems, and electronic warfare, due to their advantages in terms of selectivity, sensitivity, and rejection of unwanted signals.

However, it is important to note that heterodyne receivers require more complex circuitry and additional components compared to other types of receivers, such as DCRs.

2.3.1.1 Image Frequency

Heterodyne receivers encounter a significant challenge known as the "image problem." This arises because the mixer, which acts as an analog multiplier, does not preserve the polarity of the difference between the two input frequencies. Consequently, both positive and negative differences between the input frequencies produce the same IF. This implies that regardless of whether the input frequency is higher or lower than the LO frequency, it is translated to the same IF. In Figure 2.4, it can be observed that two spectra positioned symmetrically around the LO frequency are downconverted to the IF. The image frequency is an undesirable frequency that is symmetrically situated around the LO frequency. It is determined by adding or subtracting twice the IF from the LO frequency.

The presence of the image frequency poses a problem where two distinct RF signals, one at the desired frequency and another at the image frequency, generate the same IF following the mixing process. This can result in interference because the receiver is unable to differentiate between the desired signal and the image signal. The image frequency arises due to the transmission of signals by various users across different standards, leading to the generation of numerous interferers. If an interferer happens to fall at the frequency of the image, it can corrupt the desired signal during downconversion. While each wireless standard imposes constraints on user emissions within its own frequency band, it has no control over signals in other frequency bands. As a result, the power of the image signal can

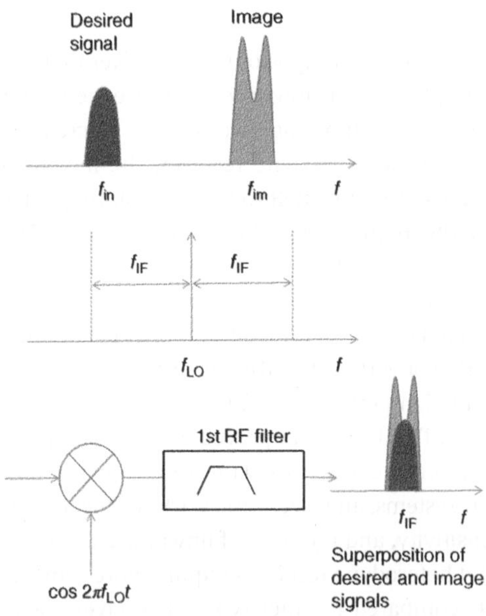

Figure 2.4 Image frequency

be considerably stronger than that of the desired signal, necessitating the implementation of effective "image rejection" techniques.

To address the issue of the image frequency, heterodyne receivers employ various techniques in their design. One commonly used approach is the inclusion of a bandpass filter at the receiver's input stage to reject signals at the image frequency. This filter acts as a barrier, preventing the image frequency signal from reaching the mixer and minimizing the potential for interference. Typically positioned before the mixer, the bandpass filter is designed to have low loss in the desired band and high attenuation in the image band. By selecting an IF that is sufficiently large, it becomes possible to satisfy both requirements simultaneously. While the front-end band-select filter can offer some degree of image rejection, it is crucial to balance its in-band loss, selectivity, and out-of-band attenuation. In practice, a filter with high image rejection is often placed between the LNA and the mixer. This configuration ensures that the gain provided by the LNA compensates for the filter's contribution to the receiver's noise figure. Achieving the necessary linearity and selectivity for the image-reject filter typically requires the use of passive, off-chip implementations operating at high frequencies. These filters are specifically designed to provide 50-Ω input and output impedances.

2.3.1.2 Image Rejection

Figure 2.5 illustrates that the frequency difference between the desired channel and the image is $2 \times f_{IF}$. Consequently, to achieve optimal image rejection, it is preferable to choose a large value for $2 \times f_{IF}$, which results in a substantial difference between f_{RF} and f_{LO}. However, it is important to consider that in a heterodyne architecture, the goal is to lower the center frequency to a level where practical filters can be used for channel selection. As $2 \times f_{IF}$ increases, the center of the downconverted channel (f_{IF}) also increases, requiring a higher Q in the IF filter. Figure 2.5 presents two scenarios that demonstrate the trade-off between high and low values of IF. A high IF allows for significant image rejection, while a low IF

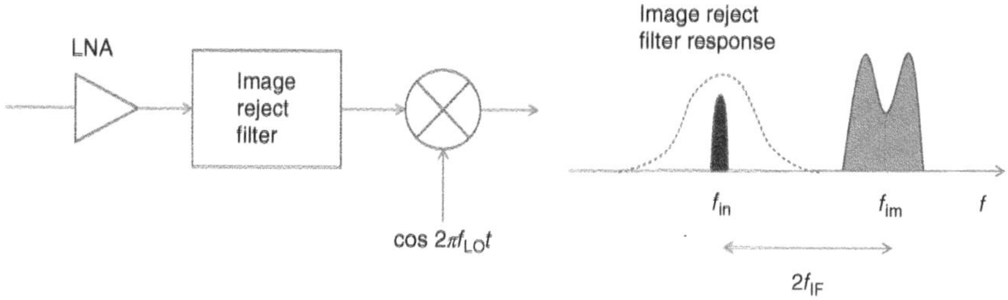

Figure 2.5 Image rejection filter

aids in the suppression of in-band interferers. Thus, it can be concluded that heterodyne receivers face a trade-off between image rejection and channel selection.

2.3.1.3 Double-Conversion Receivers

In the simplified heterodyne architecture depicted in Figure 2.5, there is often a significant trade-off between image rejection and channel selection. When the IF frequency is set to a high value, the image can be effectively suppressed, but achieving precise channel selection becomes challenging, and vice versa. To address this trade-off, the concept of heterodyning can be extended to incorporate multiple downconversion stages, known as the superheterodyne approach. Each downconversion stage is followed by filtering and amplification, allowing for partial channel selection at progressively lower center frequencies. This approach reduces the required quality factor (Q) of each filter. However, the introduction of additional downconversion steps also introduces mixing spurs, which can pose challenges in their management.

To strike a balance, most heterodyne receivers employ a double-conversion design, which includes two stages of frequency mixing, as depicted in Figure 2.1. In this configuration, the first mixing stage converts the incoming RF signal to a lower frequency known as the first intermediate frequency (IF_1). Subsequently, the second mixing stage further downconverts the signal to an even lower frequency called the second intermediate frequency (IF_2). This multistage conversion process helps effectively suppress the image frequency.

In practical applications, mixers within heterodyne receivers do not operate perfectly and can introduce undesirable effects into the receive path. One such effect is the occurrence of mixing spurs, which result from the multiplication of the RF input by the various harmonics of the LO. Managing these mixing spurs requires careful consideration of the LO frequencies chosen. Although it is technically feasible to incorporate more than two downconversion steps in the receiver design, doing so adds complexity and exacerbates the challenge of effectively managing the mixing spurs. Therefore, in most cases, it is common to limit the number of downconversion steps to maintain a manageable and practical design.

Despite the increased complexity, higher current consumption, larger size, and higher component count, the double-conversion approach provides superior performance in terms of sensitivity and selectivity. This approach has been utilized in high-performance and reliable communication devices and systems for many decades, such as pagers, analog cellular phones, and cellular base stations. Integrated circuits (ICs) like the Motorola FM Communications Receiver Integrated Circuit MC13136 [14] are still in use in the market. This single-chip IC incorporates dual conversion for FM communications, covering the entire signal path from the antenna to audio output. Typical applications include walkie-talkies, cordless telephones, and short-range communication devices.

Figure 2.6 displays the application circuit for traditional, single-channel, and narrow-band FM applications. It is evident that all the necessary building blocks for double conversion mentioned earlier are implemented within the chip, with the exception of the IF filters. The first IF frequency is 10 MHz, and the second IF frequency is 455 KHz. According to their respective datasheets [15], these ceramic IF filters offer superior frequency selectivity compared to on-chip filters. The use of ceramic materials contributes to their high-frequency selectivity, resulting in excellent image rejection and channel rejection, despite their larger physical size [16].

Achieving high-tier performance in systems often involves trade-offs, including increased lineup complexity, cost, current consumption, component count, and physical size. The double-conversion architecture, while offering superior performance, may pose challenges in terms of integration due to the need for numerous chip connections and external lumped elements.

2.3.2 Direct Conversion Receiver

The previously discussed double-conversion receiver has some drawbacks, including the use of bulky off-chip filters and two LOs, which make it difficult and expensive to use. In contrast, a DCR, also known as a zero-IF receiver or homodyne receiver, simplifies the design significantly. It directly converts the incoming radio frequency signal to baseband frequencies without the need for an IF stage. The DCR operates in a similar way to the superheterodyne receiver (SHR), but without an IF stage. Therefore, the same analysis that applies to the SHR also applies to the DCR. However, the DCR only provides protection against in-band interferers through the I-LPF (low-pass filter) and Q-LPF. Unlike the SHR, the DCR no longer has a fixed LO in the backend. Instead, the LO must be set at the same receive frequency (f_R) as the desired signal. This configuration allows the desired RF signal to be shifted directly to baseband, resulting in a zero-IF receiver.

Figure 2.7 illustrates the conversion process, highlighting a significant advantage of direct conversion. This method eliminates the need for relatively large IF filters. However, to extract information from each of the sidebands, both the I and Q mixer outputs (shown in Figure 2.8) are required, as the two sides of the RF spectrum carry distinct information. Failure to implement a scheme to extract this information would lead to irreversible corruption. The DCR lineup offers a simple and integrated design, making it well-suited for integration purposes. It has fewer components, resulting in lower costs and reduced current consumption compared to the SHR. Unlike the SHR, the operating frequency of the backend chip in the DCR is variable. This flexibility enables it to cover the entire operating range and deliver excellent performance. As a result, DCRs generally fall into the mid-tier category and must meet more stringent requirements compared to SHR receivers in order to achieve high-tier performance.

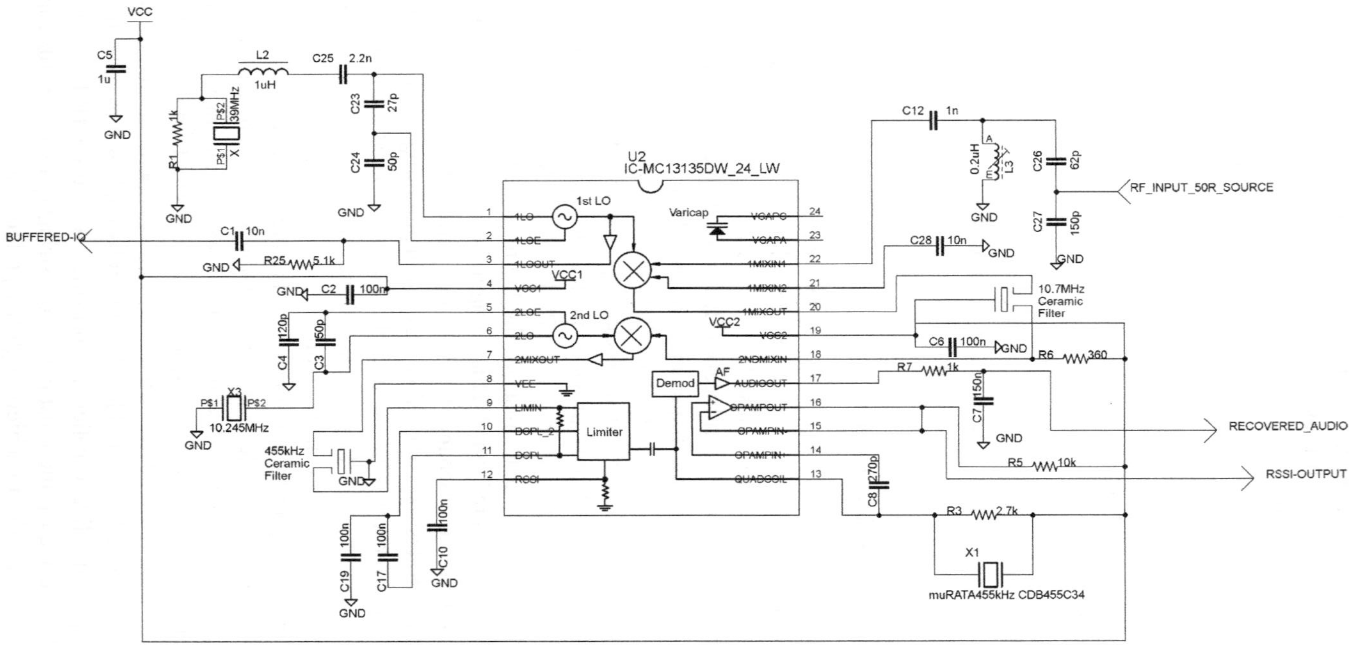

Figure 2.6 Application circuits of MC13136

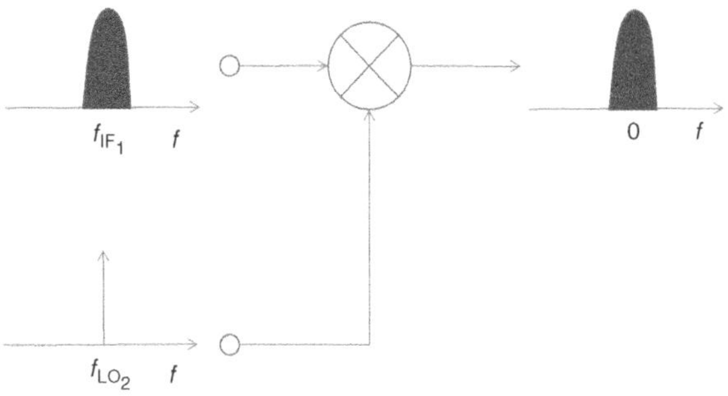

Figure 2.7 Direct conversion process

Figure 2.8 Quadrature downconversion

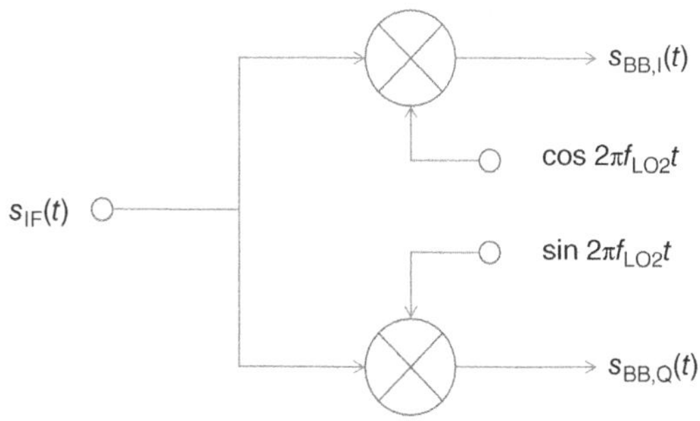

2.3.2.1 Disadvantages of Direct Conversion Receiver

While DCRs offer numerous advantages, they also exhibit certain nonidealities that can affect their performance. The most notable nonideality is the presence of DC offset in the downconverted spectrum, which stems from the self-mixing or rectification of the LO signal. In fact, the DC offset can exceed the magnitude of the downconverted signal itself, posing limitations on the receiver's dynamic range and the modulation formats it can accommodate. Addressing the DC offset issue often involves implementing high-pass filtering, which mitigates the problem but sacrifices information in the signal spectrum. This can be particularly problematic in cellular wireless systems where spectral efficiency is crucial and the channel's spectrum remains relatively constant.

In addition to the DC offset, zero-IF conversion introduces other nonidealities that can degrade receiver performance. Spurious LO leakage can occur due to reverse leakage through the RF path, leading to LO retransmission. Interferer leakage may also arise when a strong RF interferer leaks through the RF amplifier and enters the mixer via both the LO and RF ports, resulting in DC offset. Compared to heterodyne receivers, DCRs are more susceptible to undesired signals and distortion caused by second-order nonlinearities. LO self-mixing can generate a DC signal in the mixer output, potentially desensitizing or saturating the baseband amplifier, while LO frequency errors can cause asymmetric RF signal conversion around DC. Furthermore, I/Q mismatches can introduce DC offsets that exhibit minimal variation over time, necessitating analog or digital calibration techniques for their removal.

To mitigate these nonidealities, designers of zero-IF receivers must carefully balance performance, complexity, and power consumption trade-offs. Various techniques can be employed, including the use of differential LOs, multiple RF amplifier stages, balanced RF circuits, and analog or digital calibration methods. The selection of a specific technique depends on the application's requirements and the available design resources.

2.3.2.2 Design Examples of Direct Conversion Receiver

Multiple DCR ICs are available in the market, one of which is the CMX994 family offered by CML [17]. Figure 2.9 illustrates the architecture of this family, comprising the CMX994, CMX994A, CMX994E, and CMX994G ICs. These devices are DCR ICs that employ highly linear I/Q demodulators to achieve direct conversion to zero IF, near-zero IF, and low IF. The CMX994 family features flexible LO inputs, an integer-N PLL, and an on-chip negative resistance amplifier that, when combined with external components, enables the provision of a voltage-controlled oscillator (VCO).

The receiver incorporates a fully integrated LNA before the downconverter section, which can be configured with an output impedance of either 100 or 50 Ω. The 50 Ω mode provides higher gain but consumes an additional 2 mA of current, and it is primarily intended for frequencies of 450 MHz or higher. The downconverting

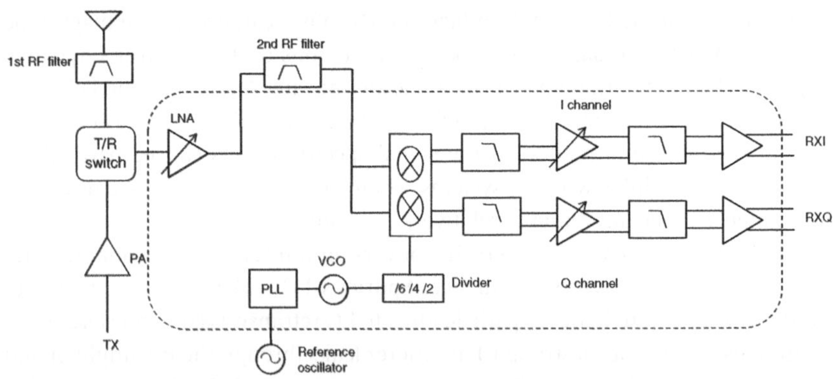

Figure 2.9 Block diagram of CMX994

mixers, known for their high linearity, are immediately followed by a baseband filter stage, where the bandwidth is determined by external capacitors. After these filters, gain is applied through a variable gain amplifier, and further filtering occurs, again controlled by external capacitors. To calibrate the internal filter circuits and ensure accurate control of the filters' cut-off point, a reference resistor must be installed.

The CMX994 family generates differential I/Q signals as its output, which can be fed into analog-to-digital converters. The receiver's I/Q chain provides a feature to correct inherent DC offsets in the hardware, optimizing the system's dynamic range. This correction is controlled by the microprocessor or DSP responsible for processing the I/Q signals. The LO section includes an integer-N PLL that can be used with the on-chip VCO or an external VCO. The on-chip VCO comprises a negative resistance amplifier and buffers, allowing external components to determine the operating frequency and tuning range. The Rx LO signal can be divided by 2, 4, or 6, and there is also a Tx LO output available, which can be divided by 1, 2, 4, or 6. Alternatively, the on-chip PLL and VCO can be disabled, and an external LO source can be utilized.

All features of the CMX994 family can be controlled through the C-BUS control interface. Furthermore, the receiver sections have a low power mode that reduces current consumption, which can be utilized when reduced IM performance is acceptable. The CMX994 family also boasts very high I/Q mixer IIP2 performance, minimizing DC offsets that are commonly encountered in DCRs. The CMX994 family incorporates digitally controlled DC offset correction, capable of reducing the offset to 25 mV or less for errors of up to $\pm$200 mV for the CMX994 and up to $\pm$800 mV for the CMX994A/CMX994E. This reduction in offset represents a negligible decrease in dynamic range, approximately 0.1 dB, for a typical ADC input signal range of 2Vp-p. The required correction must be measured externally and applied near the beginning of the I/Q baseband chain to maximize the analog sections' dynamic range.

The chip combines the advantages of DCR with minimal external components, offering versatility and programmability. Its applications span various wireless domains, including analogue/digital multimode radio, data telemetry modems, satellite communications, and software-defined radio (SDR). Chapter 4 will provide a more comprehensive exploration of the SDR design.

2.3.3 Image-Reject Receivers

The examination of heterodyne and DCRs has revealed a range of advantages and disadvantages associated with each approach. Heterodyning, while offering benefits such as image rejection, can face challenges related to image and mixing spurs. On the other hand, DCRs encounter issues with even-order distortion and flicker noise [2]. However, there is a category of receivers known as "image-reject" architectures [18, 19] that effectively eliminate the image without the need for extensive filtering. This approach allows for the avoidance of the trade-off between image rejection and channel selection. An example of an image-reject architecture is the low-IF receiver, which utilizes single-stage heterodyne mixing to convert the modulated RF carrier to a frequency slightly above DC. The specific frequency typically ranges from a few hundred kilohertz to a few megahertz, depending on the bandwidth of the RF channel. By employing this approach, the low-IF receiver successfully circumvents the DC offset issue commonly encountered in DCRs. One notable advantage of the low-IF receiver is its compatibility with higher-order modulation formats such as 8PSK and higher. However, it does require a higher-performance analog-to-digital converter compared to a DCR. This requirement is due to the need for accurate processing and conversion of the IF signal in the low-IF receiver.

The Low-IF Image-Reject Receiver operates through a series of steps to extract the desired information from the received RF signal. First, the RF signal is down-converted by mixing it with a LO signal in a mixer. The LO frequency is carefully chosen to produce the desired IF frequency when combined with the RF signal. As a result of this mixing process, the output of the mixer contains both the desired IF signal and an unwanted signal known as the image frequency.

To remove the image frequency, the mixed signal is passed through a bandpass filter specifically designed to have a narrow bandwidth centered around the desired IF frequency. This filter is crucial in attenuating the image frequency while allowing the desired IF signal to pass through with minimal distortion. By effectively rejecting the image frequency, the receiver can focus on processing the desired signal more accurately. Once the image rejection filtering is complete, the signal may undergo additional stages of processing depending on the specific requirements of the communication system. These stages may include amplification to boost the signal strength, demodulation to recover the original information, and decoding to interpret and extract the relevant data. These additional processing steps are tailored to the specific communication protocol or application, ensuring that the received signal is properly converted into usable information.

2.3.3.1 Image Rejection Architectures

Image rejection architectures are well-known and widely used for implementing structures that eliminate image interference. These architectures can be classified into two primary types: half-complex and full-complex architectures [20, 21]. Figure 2.10 provides an illustration of the half-complex and full-complex receiver architectures, which utilize polyphase filters. The half-complex architecture combines image-reject mixers with passive or active filters. In this configuration, a real RF signal is mixed with a LO complex signal to generate the input for the IF

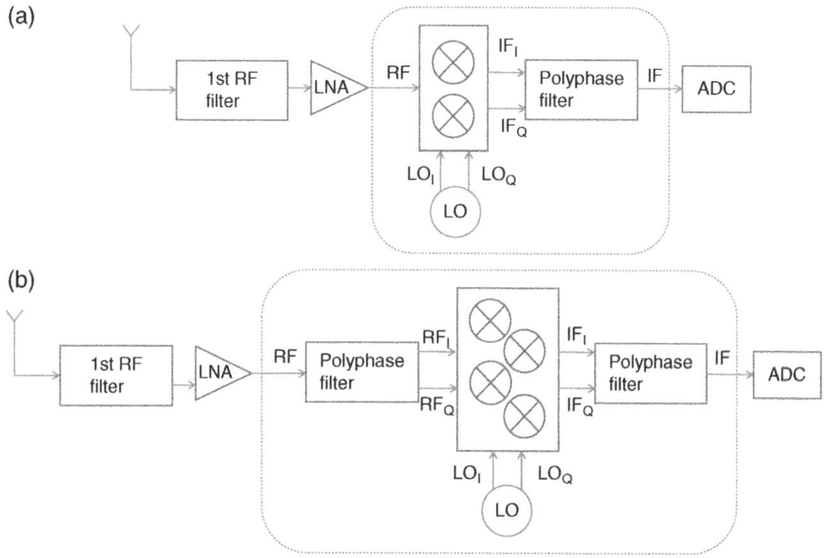

Figure 2.10 (a) Half-complex (b) and full-complex receiver architectures

polyphase filter (Figure 2.10a). The effectiveness of image rejection in this architecture relies on three key parameters:

a) The balance between the I and Q signals, including phase and magnitude errors.
b) The proper matching of mixers.
c) The performance of the polyphase filter.

The Hartley and Weaver architectures, depicted in Figure 2.11, are two commonly used architectures that employ these techniques [22, 23]. Generally, the Weaver topology is favored over the Hartley architecture. This preference arises from the fact that the Hartley architecture's 90° phase shifter block (Figure 2.11a) imposes stringent design constraints, particularly regarding component matching. As a result, significant phase errors can occur, especially at higher frequencies. For instance, even a minor 20% variation in resistors and capacitors used for quadrature generation, due to temperature or process variations, can lead to an image rejection ratio as low as 20 dB [24].

2.3.3.2 Complex Polyphase Filters

The full-complex architecture, also referred to as double quadrature downconversion, relies on the utilization of complex polyphase filters [25]. These filters are well-suited for high-frequency applications as they can meet the requirements for dynamic range and bandwidth in RF frequencies [26]. In this configuration,

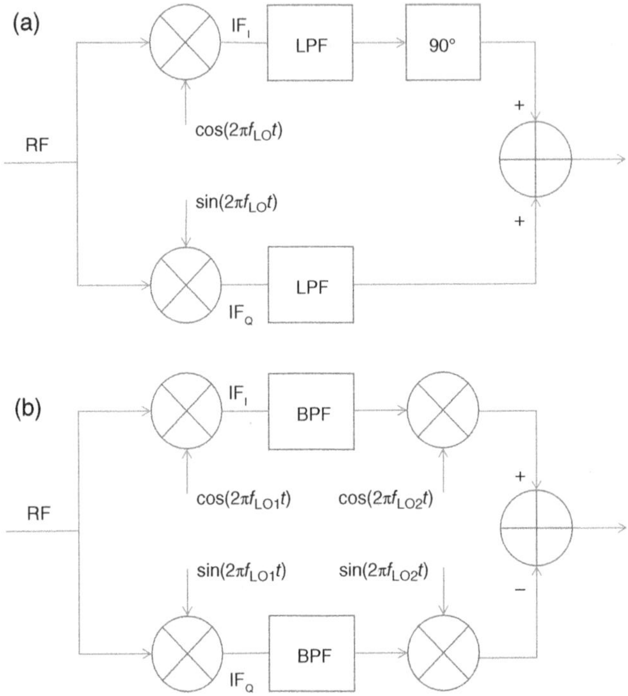

Figure 2.11 (a) Hartley and (b) Weaver image-reject architectures

instead of employing bandpass filtering, a notch frequency positioned at the image frequency is utilized to reject image signals. Figure 2.10.b illustrates the process, where the RF signal undergoes complex filtering using an RF polyphase filter. The RF and LO complex signals are then multiplied and fed into the IF polyphase filter, resulting in an image suppression of approximately 60 dB. One advantage of this design is that image rejection is supported in the RF domain by the RF polyphase filter and the quadrature LO, which is beneficial when compared to the half-complex architecture. Consequently, the design constraints for achieving image rejection are more relaxed in the RF polyphase filter and the LO, compared to the IF polyphase filter.

2.3.3.3 Disadvantages of Low-IF Image-Reject Receiver

The Low-IF Image-Reject Receiver architecture offers a range of advantages. Firstly, it effectively rejects the image frequency without the need for complex high-frequency RF filters, simplifying the receiver design process. Secondly, it reduces image interference, enabling the utilization of higher-order modulation formats. Lastly, it can be implemented using a lower-cost analog-to-digital converter compared to DCRs, as the IF frequency is typically lower. However, there are challenges associated with the Low-IF Image-Reject Receiver. Designing and optimizing the bandpass filter to achieve sufficient image rejection while maintaining the desired signal bandwidth can be a complex task. Additionally, more advanced signal processing algorithms may be required to handle nonlinearities and imperfections introduced by the mixer stage. Overall, the Low-IF Image-Reject Receiver architecture strikes a balance between image rejection performance and system complexity, making it a popular choice in various communication systems.

References

1 Vizmuller, P. (1995). *RF Design Guide: Systems, Circuits, and Equations*. Artech House.

2 Behzad, R. (2012). *RF Microelectronics*, 2e. Pearson Education, Inc.

3 Luzzatto, A. and Haridim, M. (2017). *Wireless Transceiver Design*, 2e. John Wiley & Sons.

4 Davis, W.A. and Agarwal, K.K. (2001). *Radio Frequency Circuit Design*. BWSTM.

5 Gonzalez, G. (1997). *Microwave Transistor Amplifiers – Analysis and Design*. John Wiley & Sons.

6 Lau, H. (1995). "Performance concerns guide the design of pager antennas," *Microwaves & RF*, pp. 95–106.

7 Lee, W.C.Y. (1995). *Mobile Cellular Telecommunications: Analog and Digital Systems*, 2e. McGraw-Hill.

8 Hanes, D. et al. (2017). *IoT Fundamentals: Networking Technologies, Protocols, and Use Cases for the Internet of Things*. Cisco Press.

9 Shannon, C.E. (1948). "A mathematical theory of communication," *Bell System Technical Journal* 27: 379–423, 623-656, July October.

10 Williams, A.B. (2006). *Electronic Filter Design Handbook*. McGowan-Hill Education.

11 Lau, H. (2019). *Practical Antenna Design for Wireless Products*. Artech House.

12 Bruce Carlson, A. (2001). *Communication Systems: An Introduction to Signals and Noise in Electrical Communication*, 4e. McGraw-Hill.

13 Proakis, J.G. and Salehi, M. (2001). *Communication Systems Engineering*, 2e. Pearson.

14 Motorola, Inc. (1996). "MC13135 FM communications receivers," Product Datasheet.

15 Murata Manufacturing Co., Ltd (2011). "CERAFIL product catalog".

16 Matthaei, G., Young, L., and Jones, E.M.T. (1980). *Microwave Filters, Impedance-Matching Networks, and Coupling Structures*. Artech House.

17 CML Microsystems Plc (2020). "CMX994G direct conversion receivers," Product Datasheet.

18 Razavi, B. (2000). "Image-reject architecture for low-IF and zero-IF receivers," *IEEE Journal of Solid-State Circuits* 35 (3): 294–303.

19 Long, J. R. and Nguyen, V. N. (2004). "Image-reject receiver design using low-IF architecture," *Proceedings of the IEEE 2004 Custom Integrated Circuits Conference*, pp. 473–476.

20 Crols, J. and Steyaert, M. (1998). "Low-IF topologies for high-performance analog font ends of fully integrated receivers," *IEEE Transactions on Circuits and Systems* II: 269–282.

21 Steyaert, M., Janssens, J. et al. (2000). "A 2V CMOS cellular transceiver front end," *IEEE Journal of Solid State Circuits* 35 (12): 1895–1907.

22 Razavi, B. (1996). Challenges in portable RF transceiver design. *IEEE Circuits and Devices Magazine* 12: 12–25.

23 Xu, J., Chen, J., and Zheng, J. (2001). "Design of Weaver topology," *Electronic Letters* 37 (18): 1133–1135.

24 Maligeorgos, J. and Long, J. (2000). "A 2V 5.1-5.8 GHz image-reject receiver with wide dynamic range," *IEEE Solid-State Circuits Conference ISSCC*, pp. 322–323.

25 Haddad, F. et al. (2018). *Polyphase Filter Design Meth odology for Wireless Communication Applications*. HAL OpenScience.

26 Wu, C. and Chou, C. (2004). "A 5 GHz CMOS double quadrature receiver front-end with single-stage quadrature generator," *IEEE Journal of Solid State Circuits* 39 (4): 519–521.

3

Transmitter Design

3.1 Introduction

This chapter focuses on the design of transmitters, with an emphasis on the system as a whole and the various considerations that impact their performance. The role of the critical building blocks is explored, with the oscillator serving as the fundamental frequency source for the RF transmitter, generating a stable RF carrier signal at the desired frequency. Modulation techniques, including analog and digital modulation, are discussed, highlighting their importance in facilitating the transmission of information by converting audio, video, or data signals to a higher frequency range for effective communication. The chapter also covers upconverters, which are responsible for translating baseband signals to higher frequencies suitable for transmission. Power amplification and output filtering are examined as critical components in transmitter design to achieve the desired power level and ensure clean transmission. Other considerations, such as impedance matching and linearity requirements, are also addressed.

The chapter then explores different transmitter architectures, including direct-conversion and heterodyne transmitters. The simplicity and cost-effectiveness of direct-conversion transmitters, which directly translate the baseband signal to the desired carrier frequency, are discussed. Heterodyne transmitters, which utilize frequency conversion techniques for frequency translation to mitigate the disadvantages associated with direct-conversion transmitters, are also examined. The concept of transceiver architectures, which integrate both transmitter and receiver functionalities, is introduced. Various transceiver architectures, such as full-duplex/half-duplex and simplex architectures, are discussed, along with their advantages and challenges. The chapter provides a transmitter example to illustrate the practical implementation of a transmitter using an integrated circuit (IC), as well as a comprehensive transceiver example showcasing its functional blocks and application circuit.

Introduction to Wireless System Design: From Circuits to Web-based Applications, First Edition.
Henry Lau, Ludy Liu, and Keith Chan.

By the end of the chapter, readers will have gained a comprehensive understanding of transmitter system design, including considerations, architectures, and transceiver integration. This knowledge will serve as a foundation for exploring advanced transmitter design concepts and techniques for future system designs.

3.2 Transmitter System and Considerations

A typical transmitter block diagram is shown in Figure 3.1 [1]. Most of the architectures in the actual application are determined by the frequency of operation, modulation method, output power, and the respective regulatory requirements. However, some generalizations can be made to demonstrate the general challenges on designing an optimal transmitter.

3.2.1 Oscillator

The first building block in a transmitter is the oscillator, which is modulated by a modulating signal. Traditional modulation methods [2, 3] include analog modulation techniques such as amplitude modulation (AM) and frequency modulation (FM). In an FM transmitter, the oscillator itself is modulated, a method known as direct modulation. On the other hand, in AM applications, linear amplifier stages are utilized throughout the chain to modulate the oscillator. Another more efficient AM method involves modulating the DC supply of the final device, which can be a class-C amplifier stage [2], to achieve AM. In this case, the modulator must be an audio power amplifier (PA) capable of supplying the necessary current for the high-power RF stage. Modern transmitters commonly employ more sophisticated modulation methods, such as digital modulation (to be discussed).

The oscillator serves as the fundamental frequency source for the RF transmitter, generating a stable RF carrier signal at the desired frequency. It ensures that

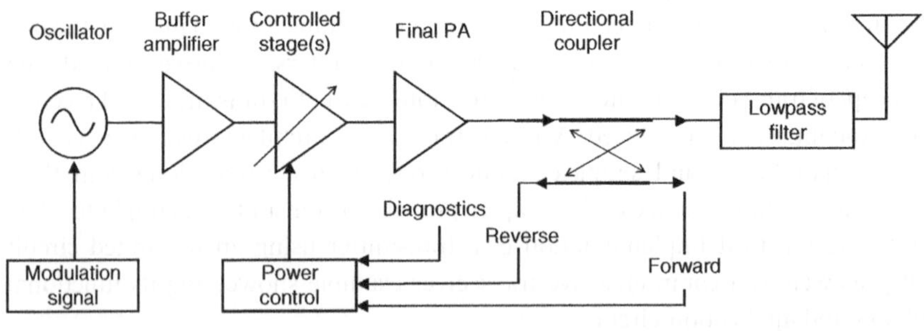

Figure 3.1 A typical transmitter block diagram

the transmitter operates at the intended frequency with precision and stability. The principle of operation for an RF oscillator involves utilizing positive feedback to maintain oscillations at the desired frequency. Positive feedback entails routing a portion of the output signal back to the oscillator's input, reinforcing and sustaining its amplitude and frequency. The specific method of achieving feedback varies depending on the type of oscillator circuit used.

There are several commonly used types of RF oscillators [2]:

a) LC oscillator: This type utilizes an inductor (L) and a capacitor (C) in a tank circuit to generate oscillations. The frequency of oscillation is determined by the values of the inductor and capacitor. However, the stability and accuracy of the generated frequency may not be optimal.
b) Crystal oscillator: Crystal oscillators utilize the piezoelectric properties of a quartz crystal to generate a highly precise and stable frequency. The oscillation frequency is determined by the natural resonance frequency of the crystal. The resonance frequency of a typical crystal can reach up to ultra high frequencies (UHFs), but with decreasing accuracy and stability.
c) Voltage-controlled oscillator (VCO): VCOs generate a frequency that can be controlled by an input voltage. They find wide application in frequency synthesizers, phase-locked loops (PLLs), and FM systems.
d) PLL: PLLs consist of a VCO, a phase detector, and a feedback loop. They synchronize the output frequency with a reference signal by comparing the phase difference between the two signals. PLLs offer versatility, accuracy, synchronization, and noise reduction capabilities, making them widely used components in various electronic systems and applications.

RF oscillators can be implemented using discrete components or ICs. The choice of oscillator circuit depends on factors such as frequency stability, tuning range, power requirements, and phase noise [4, 5] performance required for a specific application.

3.2.2 Modulation and Modulators

Modulation [2, 3] facilitates the transmission of information by converting audio, video, or data signals to a higher frequency range for effective communication. During modulation, the modulating signal's bandwidth is significantly narrower than the carrier frequency being modulated. It involves modifying specific parameters of the carrier signal, such as amplitude, frequency, or phase, based on the characteristics of the modulating signal. This modification prepares the signal for transmission over an appropriate radio channel. In more complex modulation techniques, the modulating signals may undergo digitization and encoding. Communication systems employ various modulation schemes that cater to specific requirements, including transmitting power, signal quality, frequency bandwidth, power consumption, system complexity, and cost. These schemes can be broadly classified as analog or digital modulation techniques.

3.2.2.1 Analog Modulation

Analog modulation involves the alteration of an analog carrier signal by adjusting its characteristics, including amplitude, frequency, or phase, to correspond with the transmitted information. The modulating signal, which carries the information, is typically an analog signal such as audio or video. The most common forms of analog modulation are AM, FM, and phase modulation (PM). These methods enable the encoding and transmission of analog signals across various communication systems.

Amplitude Modulation AM is a widely used modulation technique in communication systems for transmitting information by varying the amplitude of a carrier wave. It is a well-established and straightforward method of signal modulation. In AM, the amplitude of the carrier wave is adjusted in proportion to the instantaneous amplitude of the modulating signal, typically an audio signal. This modulation process involves multiplying the carrier wave's amplitude by the modulating signal, resulting in a waveform that contains the carrier frequency, a lower sideband, and an upper sideband. The lower sideband represents the carrier frequency subtracted by the modulating signal's frequency, while the upper sideband represents the carrier frequency added to the modulating signal's frequency. These sidebands carry the transmitted information and can be demodulated at the receiver to recover the original signal.

AM has found widespread use in broadcasting, particularly in AM radio, for the transmission of audio signals. However, it is susceptible to noise and interference, which can degrade the signal quality. As a result, other modulation techniques such as FM and digital modulation methods have gained popularity in modern communication systems due to their superior performance and robustness in combating noise and interference.

FM Modulation FM is a modulation technique utilized in communication systems to transmit information by modifying the frequency of a carrier wave. In FM, the frequency of the carrier wave is adjusted based on the amplitude of the modulating signal. To generate an FM signal, the carrier wave's frequency is continuously altered in response to the modulating signal. The frequency variations correspond to the instantaneous amplitude of the modulating signal, meaning that as the amplitude changes, the carrier wave's frequency deviation increases or decreases accordingly. FM modulation offers several advantages over other modulation techniques. It provides improved resistance to noise and interference compared to AM.

FM signals are less susceptible to external disturbances, making FM suitable for applications that require high-quality audio transmission and reliable signal reception. At the receiver end, FM signals are demodulated by extracting the frequency variations and converting them back into the original modulating signal. This is typically achieved using a frequency discriminator or a PLL circuit, which accurately recovers the modulating signal.

FM modulation finds wide application in various communication systems, including FM radio broadcasting, two-way radio communication, wireless microphones, and specific satellite communication setups. Its excellent audio fidelity and robustness against noise and interference make it an ideal choice for applications where maintaining signal quality is crucial. In summary, FM is a modulation technique that adjusts the frequency of a carrier wave to transmit information. Its advantages include resistance to noise and interference, making it suitable for high-quality audio transmission. FM modulation is widely used in diverse communication systems that require reliable signal reception and fidelity.

Phase Modulation PM is a modulation technique that manipulates the phase of a carrier signal to convey information. Unlike AM or FM which modify the amplitude or frequency of the carrier signal, PM focuses on altering the phase. In PM, the amplitude and frequency of the carrier signal remain constant. The modulation process involves changing the phase of the carrier signal based on the instantaneous amplitude of the modulating signal, which carries the desired information. By combining the carrier signal with the modulating signal, the phase of the carrier is adjusted proportionally to the amplitude of the modulating signal.

At the receiving end, the phase-modulated signal can be demodulated by comparing it with a reference signal. This allows for the extraction of the original information encoded in the phase variations. PM is utilized in both analog and digital communication systems. It offers benefits such as efficient use of bandwidth, resilience to noise interference, and compatibility with other modulation schemes like FM and AM. In digital communication, PM plays a crucial role in modulation schemes such as phase shift keying (PSK) and quadrature phase shift keying (QPSK), which are extensively used in wireless communication, satellite communication, and digital data transmission.

3.2.2.2 Digital Modulation

Digital modulation involves modifying a carrier signal to transmit digital information. It entails encoding binary data (0s and 1s) into specific changes in the carrier signal's amplitude, frequency, or phase to facilitate transmission across a communication channel.

Digital modulation techniques are extensively used in modern communication systems to achieve efficient and reliable data transmission. Several common digital modulation schemes include amplitude shift keying (ASK), frequency shift keying (FSK), PSK, and quadrature amplitude modulation (QAM).

Amplitude Shift Keying In ASK, the amplitude of the carrier wave is switched between two or more discrete levels to represent binary data. In its simplest form, ASK utilizes two amplitude levels: one to represent a binary "1" and another to represent a binary "0". The carrier signal's amplitude is adjusted based on the binary data being transmitted. When a "1" is to be transmitted, the carrier amplitude is set to a higher level, while for a "0", it is set to a lower level.

ASK modulation is relatively simple to implement and is often used in applications where simplicity and bandwidth efficiency are important. However, ASK is more susceptible to noise and interference compared to other modulation techniques such as PSK or FSK. ASK is commonly employed in applications such as wireless remote controls, Radio-Frequency Identification (RFID) systems, and some optical communication systems.

Frequency Shift Keying In FSK, different frequencies are used to represent different digital symbols or bits. In its simplest form, FSK uses two discrete frequencies: one to represent a binary "1" and another to represent a binary "0". The carrier signal's frequency is switched between these two frequencies based on the binary data being transmitted. For example, a higher frequency might be used to represent a "1," while a lower frequency represents a "0."

At the receiver end, the modulated FSK signal is demodulated to recover the transmitted digital data. This is typically achieved by using a frequency discriminator or a PLL circuit that detects and compares the received signal's frequency with predetermined thresholds. The frequency deviation from the carrier frequencies is used to determine the transmitted digital symbols.

FSK modulation offers advantages such as good noise immunity and relatively simple implementation. However, it requires a wider bandwidth compared to other modulation techniques like PSK or ASK to achieve the same data rate. FSK is commonly used in applications such as digital communication systems, wireless data transmission, and wireless remote controls. There are variations of FSK, including Minimum Shift Keying (MSK), which minimizes the frequency transitions to improve spectral efficiency, and Gaussian Frequency Shift Keying (GFSK), which uses a Gaussian filter to shape the frequency transitions for better spectral characteristics.

Phase Shift Keying PSK is a digital modulation technique used in communication systems to encode digital symbols by altering the phase of a carrier signal. The commonly used PSK variants include Binary Phase Shift Keying (BPSK), QPSK, and higher-order PSK schemes like 8-PSK or 16-PSK.

BPSK employs two distinct phases, typically 0° and 180°, to represent binary "0" and "1" symbols, respectively. The carrier signal's phase is switched between these two values based on the transmitted digital data. QPSK, on the other hand, utilizes four different phases, usually 0°, 90°, 180°, and 270°, to represent two bits per symbol. Each symbol corresponds to a combination of two bits, doubling the data rate compared to BPSK. Higher-order PSK schemes, such as 8-PSK or 16-PSK, employ a greater number of phases to encode more bits per symbol. For instance, 8-PSK can represent three bits per symbol using eight distinct phases.

PSK modulation offers advantages such as efficient bandwidth utilization and resilience against noise. However, it is more susceptible to phase errors and necessitates more complex demodulation methods compared to simpler modulation techniques like ASK or FSK. PSK is widely employed in digital communication systems, encompassing wireless communication, satellite communication, and data transmission applications.

Quadrature Amplitude Modulation In QAM, the amplitude and phase of the carrier signal are simultaneously modified to represent digital symbols. The constellation diagram is commonly used to visualize the different combinations of amplitude and phase that correspond to specific symbols. The most common form of QAM is QAM with a square constellation (QAM-16), where there are 16 possible combinations of amplitude and phase. This allows the transmission of four bits per symbol. Higher-order QAM schemes, such as QAM-64 or QAM-256, use larger constellations with more combinations, enabling the transmission of more bits per symbol.

At the receiver end, the modulated QAM signal is demodulated to recover the transmitted digital data. This involves determining the amplitude and phase of the received signal and mapping it back to the corresponding digital symbols. QAM offers several advantages, including high spectral efficiency, as it allows the transmission of multiple bits per symbol. It is widely utilized in various communication systems, including digital TV, wireless communication (such as Wi-Fi and 4G/5G), and cable modems. QAM also provides robustness against noise and interference, making it well-suited for high-speed data transmission applications.

3.2.3 Upconverter

An upconverter is a crucial component used in wireless communication systems to convert a baseband or intermediate frequency (IF) signal into a higher frequency suitable for transmission. Its primary function is to modulate and transmit RF signals effectively. During the upconversion process, the baseband or IF signal is combined with a local oscillator (LO) signal to generate the desired RF carrier frequency. The LO signal, derived from a stable oscillator, determines the transmission frequency. To accomplish this, the upconverter utilizes a nonlinear device, such as a mixer or multiplier, to merge the baseband/IF signal with the LO signal.

Similar to the mixer used in RF receivers, the mixer or multiplier in transmitters performs frequency translation by multiplying the baseband/IF signal with the LO signal. This multiplication generates the sum and difference frequencies of the two signals. The desired RF carrier frequency is typically obtained by selecting either the sum or difference frequency using a filter. This process is known as heterodyning. To ensure that the signal meets the necessary transmission requirements, additional circuitry for amplification, filtering, and impedance matching may be employed after the upconverter. These supplementary components work together to optimize the signal quality following the upconversion stage.

3.2.4 Power Amplifier

The PA is an essential component in an RF transmitter. Its primary function is to amplify a low-power RF signal to a higher power level, making it suitable for transmission over the airwaves. By increasing the signal strength, the PA prepares the signal for broadcast by feeding it to an antenna. The main purpose of a PA is to amplify the RF signal received from the preceding stages of the transmitter. This RF signal, which often originates from a modulator or upconverter, typically has a relatively low power level. The PA enhances the signal's power while preserving its fidelity and inherent characteristics. In addition to providing amplification, it is crucial for the PA to maintain linearity in order to preserve the integrity of the modulated RF signal. Linearity ensures that the amplitude and phase variations of the input signal are faithfully reproduced at the output without introducing distortions. This is especially critical in applications like wireless communication systems, where the transmitted signal carries important information, and any distortion can negatively impact the quality of the received signal. PAs aim to achieve high efficiency by maximizing the conversion of DC power supply into useful RF power while minimizing power losses. High-efficiency PAs have several benefits, including reduced power consumption, decreased heat dissipation and lower overall system costs. By efficiently utilizing power resources, these amplifiers contribute to improved energy efficiency and help optimize the performance and cost-effectiveness of RF systems.

The PA section may consist of the fundamental building blocks of a buffer amplifier, power-controlled stages, a power controller, a final PA, and a directional coupler in some designs. The signal level from the output of the upconverter is typically very low, so a PA is necessary to provide amplification and deliver sufficient signal power to the antenna for radiation.

Depending on the required gain and output power, a buffer amplifier (also known as a driver amplifier) may be used to serve as an intermediate stage between the low-power input signal source and the PA. The buffer amplifier provides isolation between the low-level input signal source and the PA stage. It prevents the PA from affecting the characteristics of the input signal source and vice versa. This isolation helps maintain the stability and performance of the overall system. The buffer amplifier also amplifies the low-power input signal to a level sufficient to drive the PA effectively. It ensures that the PA receives an adequately amplified signal to achieve the desired output power without distortion. The buffer amplifier also helps to improve the signal-to-noise ratio and dynamic range of the system.

In certain communication systems, it is essential to have adjustable transmit output power for optimal overall performance. In cellular communications, the output power is dynamically adjusted to ensure that only the necessary power is used, reducing current consumption and extending usage time between recharges. Setting the output power too high can overload the corresponding receiver or other receivers, leading to signal distortion and degradation in signal quality. To address this, a gain-controlled circuit block is positioned between the buffer and the PA as a control stage. This control stage not only regulates the power but also enhances isolation between the buffer amplifier and the PA, reducing the occurrence of the frequency pulling effect. The frequency pulling effect, also known as frequency pulling or frequency pushing, refers to the alteration or shift in the output frequency of an RF transmitter caused by changes in the circuit's load impedance or other influencing factors. This effect is generally undesired in RF transmitters as it results in performance degradation, decreased spectral purity, or system instability.

By incorporating the buffer amplifier and the gain-controlled circuit block, higher isolation is achieved between the VCO and the PA, minimizing the frequency pulling effect resulting from impedance changes on the PA or the antenna.

The choice of PA type in RF transmitters depends on several factors, such as the application, desired power output, frequency range, linearity requirements, efficiency, and cost considerations. Various types of PAs are commonly employed in RF transmitters. Below are examples of frequently used PA types [2]:

Class A Amplifier Class A amplifiers provide high linearity and low distortion but are relatively inefficient. They operate in a linear mode throughout the RF cycle. Class A amplifiers are suitable for applications requiring excellent linearity, such as laboratory equipment or low-power transmitters.

Class AB Amplifier Class AB amplifiers combine the linearity of Class A amplifiers with improved efficiency. They operate in a combination of linear and nonlinear modes. Class AB amplifiers are commonly utilized in applications requiring moderate power output, like mobile communication systems or small-scale broadcasting.

Class B Amplifier Class B amplifiers offer higher efficiency compared to Class A and Class AB amplifiers but introduce some distortion due to their push–pull configuration. They employ separate transistors for positive and negative halves of the RF cycle, enhancing efficiency. Class B amplifiers are used in applications where efficiency is crucial, such as RF transmitters for FM radio broadcasting.

Class C Amplifier Class C amplifiers are highly efficient but exhibit poor linearity as they operate in a highly nonlinear mode. They are suitable for applications demanding high power output and efficiency while linearity is not critical. Class C amplifiers find common usage in RF transmitters for applications like amateur radio and RF power amplification.

Class D Amplifier Class D amplifiers, also known as switching amplifiers, utilize pulse-width modulation techniques to achieve high efficiency. They function as switches rather than linear amplifiers. Class D amplifiers are commonly employed in RF transmitters for applications such as wireless communication systems and digital broadcasting.

Doherty Amplifier Doherty amplifiers combine a Class AB or Class B amplifier with a peaking amplifier. This architecture enables high efficiency by dynamically adjusting the power supply voltage to the peaking amplifier. Doherty amplifiers find extensive use in high-power RF transmitters, including cellular base stations and television broadcasting.

In high-power applications, a power-control building block is included to monitor diagnostic signals, which typically encompass various parameters such as the current of the final device, temperature, controlled stage voltage drive level, forward power levels, and reverse power levels. To effectively control and monitor these parameters, feedback loops are incorporated as part of the control block. A directional coupler can be added as part of the feedback system to monitor the forward and reverse power levels.

For certain high-power applications, the instantaneous output voltage can become very high if the output is not properly matched to the antenna or if the antenna is broken. This can result in the output voltage exceeding the rated value of the output active device, potentially causing a breakdown and malfunction of the complete transmitter. To protect the PA, a directional coupler is typically added at the output of the PA to sense the forward and backward power. The degree of mismatch can be determined from the forward and backward powers.

If a significant mismatch occurs, indicating a large deviation from optimal conditions, the PA can be shut down instantly to prevent any potential issues. By incorporating a directional coupler in the feedback system, the PA is safeguarded against potential mismatches and excessive output voltage, ensuring the overall stability and functionality of the transmitter.

In demanding applications, it is common to include an isolator between the PA and the antenna. RF isolators are two-port devices designed to provide high isolation in a single direction. They allow power to flow from the input port to the output port while blocking or attenuating power in the reverse direction. This prevents reflected power from reaching the source and safeguards the source from impedance mismatches and potential damage.

Another objective of using an isolator is to reduce intermodulation [1, 6] of the main carrier with external interferers, as depicted in Figure 3.2. These external interferers can arise from other transmitting signals or the reflected signal of the carrier itself due to nearby objects. By incorporating an isolator, the system ensures that unwanted power reflections are minimized, protecting the PA and improving overall signal quality by reducing intermodulation caused by external interferers.

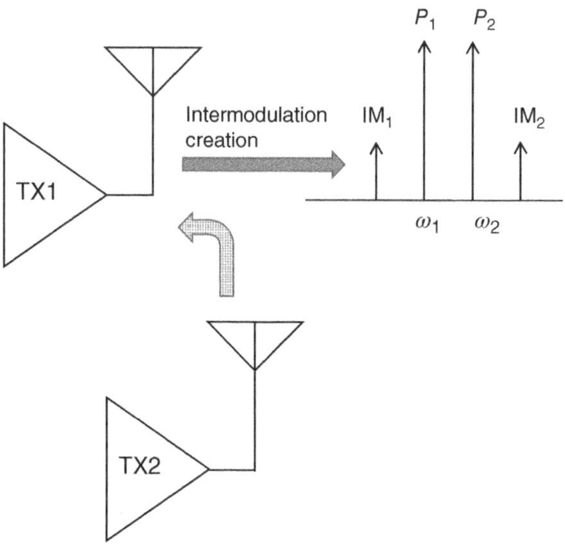

Figure 3.2 Intermodulation products caused by an external interferer

3.2.5 Output Filtering

PAs frequently utilize impedance matching networks to ensure that the input and output impedances of the amplifier are appropriately matched with the preceding and following stages of the transmitter. By properly matching the impedances, the PA can maximize the power delivered to the load and minimize the loss of signal energy due to impedance mismatches, resulting in improved overall system performance, particularly in terms of efficiency.

The low-pass filter in Figure 3.1 serves two purposes: attenuating the harmonics of the transmitted signal and providing impedance matching for the fundamental frequency. While this may appear to be a straightforward task, it is actually more complex in practical applications. Achieving the theoretical harmonic attenuation of the low-pass filter is rarely feasible, as most filters experience reflection and attenuation due to impedance mismatch at harmonic frequencies. This results in undesired harmonics being reflected back into the device and mixed again, effectively increasing the level of harmonic output from the active device and compromising the selectivity of the low-pass filter. Therefore, it is necessary to over-design the low-pass filter to provide approximately 20 dB more rejection than initially required.

3.2.6 Other Considerations

The arrangement of the directional coupler, power sensor, and low-pass filter is crucial in transmitters that operate across a wide range of output powers. When dealing with such transmitters, it is important to ensure that the directional coupler has an appropriate coupling factor for accurate sampling of the amplifier output at the lowest power setting. However, the coupling should not be excessively tight, as it can allow harmonics generated by the detector diodes to couple back into the output at the highest power setting. In certain cases, a low-pass filter might be necessary after the directional coupler to attenuate these harmonics. One drawback of this configuration is that the power-leveling loop cannot compensate for the frequency response and loss caused by the low-pass filter.

Directly placing the power sampling directional coupler at the output of a high-power stage is not recommended. The output of the coupler often contains a significant amount of harmonics, and depending on the specific implementation, these harmonics may be preferentially coupled to the detector. As a result, inaccurate power indication at the fundamental frequency can occur. In high-power PAs, diagnostic signals typically include the current of the final device, temperature, controlled stage voltage drive level, and forward and reverse power levels. These parameters are closely monitored and controlled using various feedback loops to ensure proper operation. It is crucial to exercise caution and prevent the introduction of low-frequency instability into these control loops.

To address the generation of spectral components on adjacent channels, which is particularly important in data communications where the transmitter is rapidly switched on and off during data bursts, the power-control loop is often utilized. This loop intentionally slows down the turn-on and turn-off times of the transmitter. The purpose of this deliberate slowing down is to minimize the bandwidth of the energy produced in the spectrum during the power ramp-up and ramp-down process.

The modulation circuitry faces the main challenge of maintaining consistent modulation (FM deviation or percentage AM) across a wide RF frequency range and a wide baseband frequency range. This challenge becomes more pronounced when dealing with digital signals that have significant low-frequency and DC components. In such cases, FM modulation of synthesized signal sources becomes difficult due to the corrective action of the PLL on any slowly changing frequency shift imposed on the VCO. This requires modulation of the reference frequency as well. On the other hand, rapid frequency changes in the reference source do not impact the VCO output as they do not propagate through the loop filter. Consequently, low-frequency components are directed to the reference, while high modulating frequencies are directed to the VCO in a precise balance of amplitude and phase. This ensures a constant frequency deviation for all modulating frequencies.

Load pull [7, 8] is the phenomenon where the transmitter frequency changes with increasing power output. There are four reasons for this frequency shift. Firstly, as the power output increases, changes in the device's impedance can lead to variations in the load impedance of the oscillator, resulting in a frequency shift. To address this, sufficient buffering is necessary to isolate the oscillator from load impedance variations. However, it is important to note that a buffer stage operating in amplitude compression will have reduced reverse isolation. One approach to improve isolation across a wide frequency range is to use a chain of class-A buffers with attenuators in between. Another technique to mitigate load pull in synthesized transmitters involves temporarily widening the bandwidth of the PLL during transmitter keying. By allowing a wider loop bandwidth, faster frequency correction can be applied to compensate for any frequency shift caused by changing loads.

When there is a sudden requirement for a large DC current, it can cause a momentary dip in the power supply voltage of the system. This can affect systems that lack their own DC supply regulator. VCOs, in particular, require a well-filtered and regulated power supply.

If the oscillator is not shielded from the transmitter or if RF energy is allowed to flow back to the oscillator through power supply leads, grounds, or control lines, it can disrupt the phase relationships in a PLL. This disruption occurs because the phase of the high-power signal may differ from the existing phase. In extreme cases, the high-level signal can cause rectification, leading to de-bias, and operating-point changes in the oscillator.

Increasing bias current flowing into a PA can induce voltages in coils and long circuit board traces due to changing magnetic fields. The rise time of this field is relatively long, and conventional shielding may not be effective in protecting sensitive circuitry. To achieve effective shielding, the thickness of a metal shield should be several skin depths. For example, for a rise time of 0.25 ms, the skin depth in copper is over 2 mm. Ferromagnetic (steel) shields are more effective than copper in mitigating the effects of changing magnetic fields.

3.3 Transmitter Architectures

3.3.1 Direct Conversion

Based on the receiver designs discussed in the previous chapter, most transmitter architectures follow similar topologies, but with reversed operations. For instance, if a receiver cascade includes a Low Noise Amplifier (LNA) and quadrature downconversion mixers, a corresponding transmitter cascade would involve quadrature upconversion mixers and a PA. As explained earlier, RF transmitters handle modulation, upconversion, and signal power amplification. In many modern wireless systems, quadrature baseband signals are used to transmit data. For example, the GSM system utilizes Gaussian Minimum Shift Keying (GMSK) modulation, represented by its waveform [6] as

$$x_{\mathrm{GMSK}}(t) = A(t)\cos[\omega_c t + \emptyset(t)] = A(t)\cos\omega_c t\cos[\emptyset(t)] - A(t)\sin\omega_c t\sin[\emptyset(t)] \tag{3.1}$$

To facilitate the transmission, quadrature baseband signals are defined as:

$$x_{\mathrm{baseband},I}(t) = A(t)\cos[\emptyset(t)] \tag{3.2}$$

$$x_{\mathrm{baseband},Q}(t) = A(t)\sin[\emptyset(t)] \tag{3.3}$$

As depicted in Figure 3.3, a transmitter is built utilizing these baseband signals. This particular kind of transmitter, referred to as a "direct-conversion" transmitter, employs a "quadrature upconverter" to directly convert the baseband spectrum to the RF carrier. The upconverter is subsequently followed by a PA and a matching network. The purpose of the matching network is to optimize power transmission to the antenna while suppressing out-of-band components that arise due to the nonlinearity of the PA.

It is important to note that the baseband signal, which is created within the transmitter and has a sufficiently high amplitude (several hundred millivolts), is less affected by mixer noise compared to receivers. When designing the transmitter, the process begins with the PA, which is carefully engineered to provide

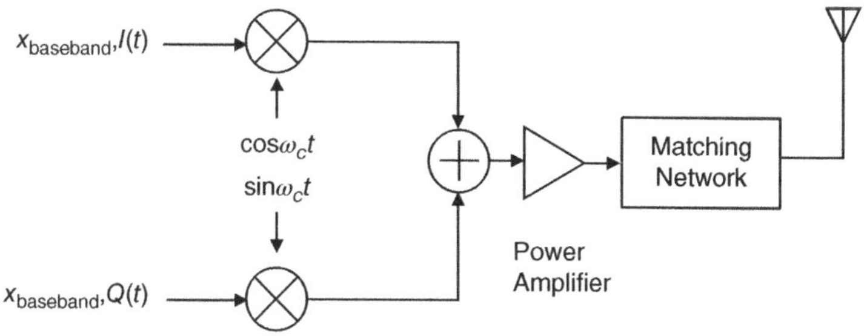

Figure 3.3 Direct-conversion transmitter

the required power to the antenna while maintaining appropriate linearity, if necessary. Since the PA has a significant input capacitance due to the use of large transistors to handle high currents, a predriver is often placed between the upconverter and the PA to serve as a buffer.

Among the transmitter architectures discussed in this chapter, the direct-conversion approach stands out as a compact and efficient solution, yielding a relatively clean output. Its advantages stem from the fact that the output spectrum primarily contains the desired signal centered around the carrier frequency, with minimal presence of undesirable spurious components. This characteristic closely aligns with that of direct-conversion receivers. In summary, some of the benefits of the direct-conversion approach are:

3.3.1.1 Simplicity

Direct-conversion transmitters provide a streamlined design compared to traditional superheterodyne transmitters. They remove the need for multiple frequency conversion stages, such as mixers, filters, and LOs. This simplification leads to reduced complexity, lower costs, decreased power consumption, and a more compact size.

3.3.1.2 Wideband Operation

They have the ability to operate over a wide frequency range. By not depending on fixed IF stages, they can potentially cover a broad spectrum without requiring additional frequency conversion stages.

3.3.1.3 Improved Phase Noise Characteristics

They exhibit superior phase noise characteristics when compared to superheterodyne transmitters. By circumventing the phase noise degradation introduced by multiple frequency conversion stages, they enhance the overall performance of the system.

Direct-conversion transmitters offer a range of benefits, as previously discussed. However, it is important to acknowledge certain challenges and limitations associated with them. These include issues such as DC offset, I/Q imbalance, image rejection, and frequency pulling:

3.3.1.4 DC Offset

The transmitter illustrated in Figure 3.3 generates quadrature signals using its analog baseband circuitry. Both the baseband circuitry and the baseband port of each upconversion mixer may exhibit direct current (DC) offsets [6, 9]. Consequently, the resulting output signal can be represented as

$$\begin{aligned} \mathrm{Vo(t)} &= [\mathrm{DC_{O1}} + A(t)\ \cos(\emptyset)]\ \cos(\omega_c t) - [\mathrm{DC_{O2}} + A(t)\ \sin(\emptyset)]\ \sin(\omega_c t) \\ &= A(t)\ \cos\ (\omega_c t + \emptyset) + \mathrm{DC_{O1}}\ \cos(\omega_c t) - \mathrm{DC_{O2}}\ \sin\ (\omega_c t) \end{aligned} \tag{3.4}$$

where $\mathrm{DC_{O1}}$ and $\mathrm{DC_{O2}}$ are the total DC offsets referred to the input port of the mixers.

The last two terms represent the unmodulated carrier part, the carrier leakage part with a relative leakage of

$$\frac{\sqrt{(\mathrm{DC}_{01}^2 + \mathrm{DC}_{02}^2)}}{\sqrt{\overline{A^2(t)}}} \tag{3.5}$$

The undesired leakage not only distorts the signal constellation, causing a higher error vector magnitude in the output signal, but it also impacts the automatic power-control circuitry when the transmitter needs to adjust the transmit output power.

3.3.1.5 I/Q Imbalance

Like direct-conversion receivers, I/Q imbalance in a transmitter leads to distortion in the constellation and degradation of the error vector magnitude. If the initial raw I/Q matching of circuits is inadequate, calibration methods can be employed. To tackle this problem, the gain and phase mismatch must be measured initially and subsequently corrected.

3.3.1.6 Frequency Pulling

The frequency pulling effect poses a significant challenge in direct-conversion transmitters. The high signal level at the PA output can result in electromagnetic coupling, where a portion of the signal can be coupled to the LO through various paths like PCB traces, inductive components such as the inductor, and the IC package on the PCB. If the coupled signal is sufficiently strong, it can pull the oscillating signal, leading to constant frequency drift or FM. Both of these effects degrade the quality of the transmitted signal. To tackle the frequency pulling issue, a more modern design approach is employed to ensure that the LO frequency is different from the PA output frequency. In this design, the frequency of the LO is set to be twice the frequency of the carrier ($\omega_{LO} = 2\omega_c$) by utilizing a divide-by-2 circuit to generate quadrature phases after the LO. This particular architecture has gained popularity for two key reasons. Firstly, it significantly reduces injection pulling. Secondly, the divider readily provides quadrature phases of the carrier, which is otherwise challenging to achieve.

3.3.2 Heterodyne Transmitters

To mitigate frequency pulling, an alternative method involves utilizing a two-step signal upconversion process, ensuring a significant separation between the LO frequency and the output spectrum of the PA. This method, depicted in Figure 3.4, bears resemblance to the "dual" architecture of heterodyne receivers discussed in the previous chapter. In this configuration, the baseband I and Q signals are initially upconverted to an IF of ω_1. Subsequently, the resulting signal is mixed with ω_2, effectively translating it to a carrier frequency of $\omega_1 + \omega_2$. To address the undesired output at $\omega_1 - \omega_2$ produced by the second mixer, a bandpass filter is applied after this stage.

Similar to the receiver counterpart, this architecture offers several advantages. One notable benefit is that the upconversion of I/Q signals occurs at a significantly lower frequency than the carrier frequency, resulting in reduced gain and phase mismatches. The equations quantifying the impact of these mismatches are the same as those derived in the previous section for direct-conversion transmitters.

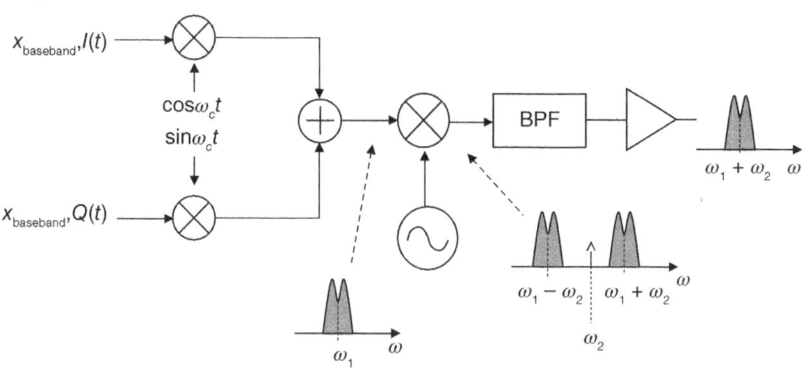

Figure 3.4 Heterodyne transmitter

Another advantage of heterodyne transmitters is their ability to effectively suppress spurious emissions, which are unwanted signals that can cause interference and disrupt communication. Through careful design of the mixing process and the use of filtering techniques, spurious emissions can be minimized, ensuring compliance with regulatory standards. Additionally, heterodyne transmitters can achieve enhanced linearity performance, particularly when employing linearization techniques such as pre-distortion or feedback mechanisms. This allows the transmitter to maintain signal integrity and reduce distortion, resulting in improved overall system performance.

3.4 Transceiver Architectures

In many wireless devices and systems, bidirectional communication is essential, requiring the integration of a receiver and a transmitter into a single transceiver unit. This transceiver unit combines the functionality of both the transmitter and the receiver, along with additional circuitry. Various transceiver architectures are utilized in conjunction with widely accepted access methods such as Frequency Division Multiple Access (FDMA), Time Division Multiple Access (TDMA), and Code Division Multiple Access (CDMA) [10, 11]. These access methods determine how multiple users can efficiently share common communication channels in the physical domain.

RF communication channels can be categorized as either paired or unpaired. Paired channels consist of two distinct RF channels operating at different frequencies, with one channel dedicated to transmission and the other for reception. On the other hand, unpaired channels utilize a single RF channel that is alternately used for transmission and reception on a time-sharing basis. Transceivers operating on paired channels, capable of simultaneous transmission and reception, are referred to as operating in full-duplex mode. Transceivers that cannot perform simultaneous transmission and reception are referred to as operating in half-duplex mode. The architectures for full-duplex and half-duplex transceivers are essentially the same. In comparison, transceivers operating on unpaired channels, known as simplex mode, have a simpler architecture compared to the duplex case.

3.4.1 Full-Duplex/Half-Duplex Architecture

In full-duplex mode, the receiver and transmitter of a transceiver can operate simultaneously on paired channels with a significant frequency separation. This simultaneous operation is made possible by utilizing a shared antenna and incorporating isolating devices known as duplexers [10, 12]. A duplexer is a three-port device that includes a transmit port, a receive port, and an antenna input-output port. It consists of interconnected bandpass filters, with each filter dedicated to either the transmit or receive band. These filters, in conjunction with the duplexer, ensure proper isolation between the transmitter and receiver. This isolation prevents the transmitter's power from causing damage to the receiver or interfering with its performance, while still allowing the antenna to function independently for both components. Consequently, a duplex radio can transmit and receive simultaneously and continuously using a single antenna.

It is crucial to understand that two duplex transceivers operating on the same paired transmit and receive frequencies cannot directly communicate with each other. In a duplex RF network, all subscriber radios, commonly referred to as mobile radios, make use of the same set of paired channels. These radios establish communication with a central base station or gateway transceiver that consistently operates in full-duplex mode. The paired channel frequencies of the base station or gateway are reversed in comparison to the subscribers. Specifically, the transmit frequency of the base station/gateway corresponds to the receive frequency of the subscribers, and vice versa.

The base station or gateway acts as a repeater, receiving data from a transmitting remote and relaying it to other listening remotes. It also manages communication between the subscribers. Full-duplex operation is crucial for the base station/gateway to perform retransmission and effectively handle the subscribers, regardless of whether they are currently transmitting or not. In this context, the transmit frequency of a subscriber is known as the uplink, typically lower than the corresponding receive frequency referred to as the downlink.

A typical implementation of a duplex subscriber architecture is depicted in Figure 3.5 [9]. The receiver adopts a heterodyne design with an IF equal to the transmit-receive frequency separation. The receiver section of the duplexer also functions as the first bandpass filter. The transmitter, likewise, follows a heterodyne approach with an IF frequency of $2 \times f_{IF}$, twice the IF frequency of the receiver. The transmitter section of the duplexer also acts as the harmonic suppression filter. By choosing these IF frequencies, a common offset LO can be utilized to generate both the uplink carrier at f_T and the injection signal to receive the downlink channel at f_R.

3.4.2 Simplex Architecture

A simplex transceiver primarily consists of a transmitter and a receiver that operate on the same channel frequency in a time-division manner. It alternates between transmitting and receiving functions. By utilizing the same frequency, $f_0 = f_T = f_R$, for both transmission and reception, subscribers can communicate directly with each other without the need for a central base station or gateway. The hardware typically incorporates a shared LO that can switch to the required

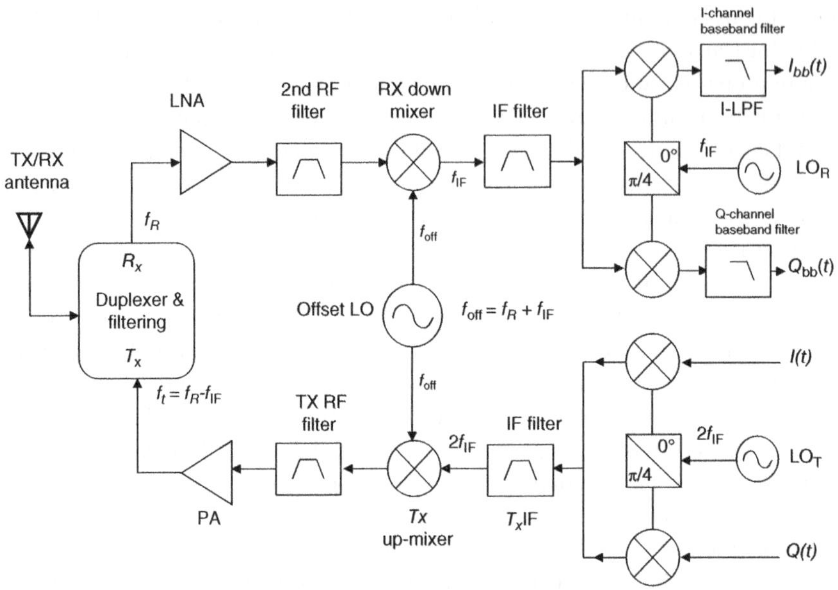

Figure 3.5 Full-duplex transceiver

frequency when necessary. Additionally, a single antenna is connected to either the active transmit or receive section through an RF switch called the antenna switch. There are various simplex approaches available that can be implemented. A simple simplex architecture comprises a direct-conversion receiver and a direct-conversion transmitter, as depicted in Figure 3.6. The transmit/receive control mechanism links the LO power and the antenna to the active section. In this configuration, the LO is accurately tuned to the channel frequency f_0 for both receiving and transmitting modes. Hence, during reception, it is crucial to take precautions to avoid LO radiation, as it can potentially interfere with nearby receivers.

3.4.3 Transmitter Example

An example of a transmitter design is the IC ADMV4530D, provided by Analog Devices [13]. It is a highly integrated upconverter specifically designed for next-generation Ka band satellite communications. The functional block diagram in Figure 3.7 illustrates the main components of the chip, including an oscillator, frequency synthesizer, I/Q upconverter mixer, and gain and variable attenuation stages.

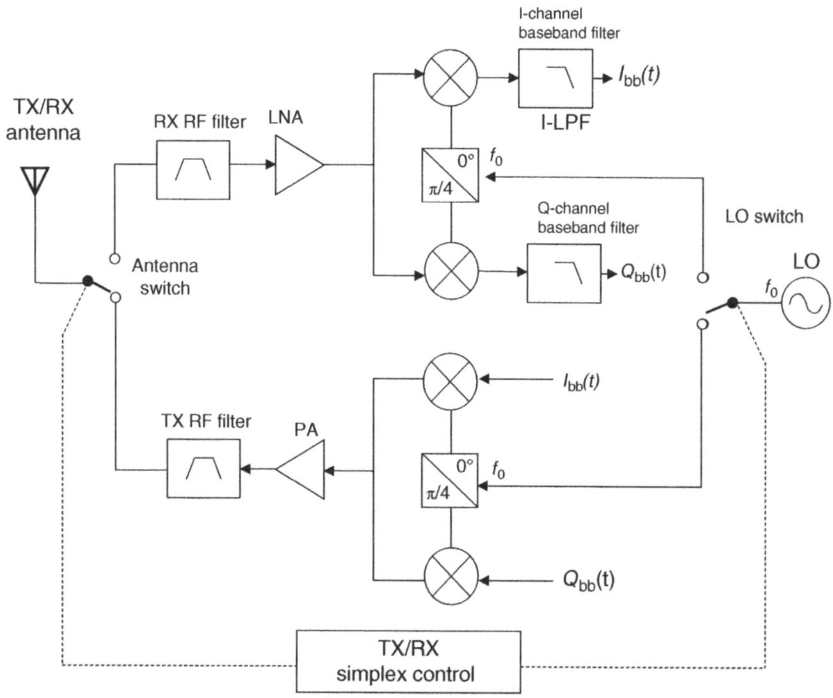

Figure 3.6 Simplex transceiver

One of the notable features of this IC is its low phase noise, fractional-N PLL, which includes a VCO and an internal 2× multiplier. The 2× multiplier enables the VCO to operate at half of the carrier frequency, effectively minimizing the frequency pulling effect discussed in the direct-conversion transmitter section. The on-chip PLL generates the required LO signal for the I/Q mixer, eliminating the need for external frequency synthesis. Additionally, the VCO incorporates an internal autocalibration routine, enabling the PLL to rapidly and accurately select the necessary settings and achieve lock-in within approximately 100 μs. This short lock-in time is particularly advantageous for applications such as frequency hopping communication systems [14, 15].

The upconverter incorporates a quadrature mixer that offers two distinct operating modes: the I/Q mode and the IF mode. In the I/Q mode, the upconverter functions as a quadrature upconverter within a direct-conversion transmitter. By applying a differential DC offset to the I and Q inputs, the LO leakage can be significantly suppressed. On the other hand, in the IF mode, the mixer serves as the secondary mixer in a heterodyne transmitter, allowing the upconversion of a

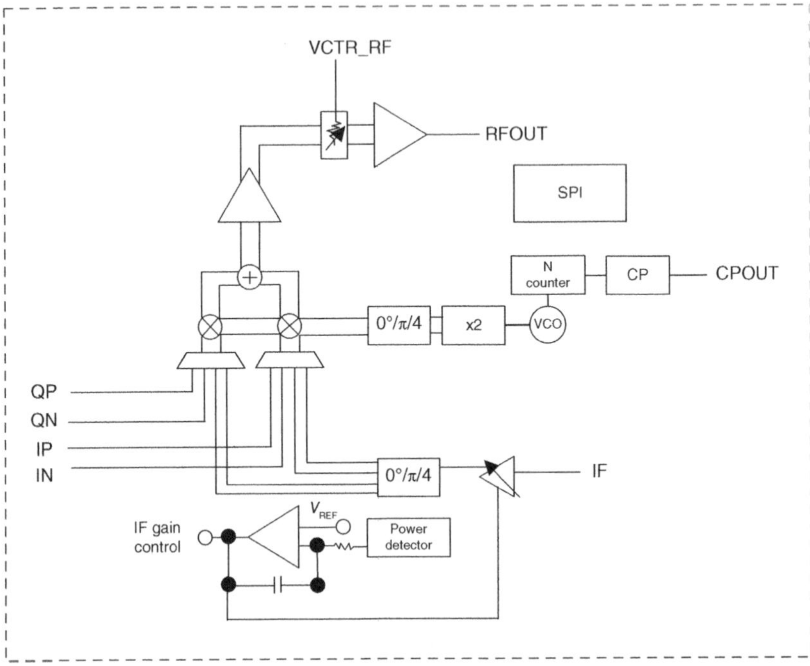

Figure 3.7 Functional block diagram of ADMV4530

real IF input ranging from 2 to 3 GHz to the carrier frequency of 27–31 GHz. This versatility ensures compatibility with various radio architectures and legacy systems.

To tackle input power variations, the upconverter is equipped with an IF automatic gain control (AGC) and an IF level detector. The AGC adjusts the IF variable gain amplifier (VGA) to compensate for changes in input power. In the functional block diagram, the RF output section of the ADMV4530 consists of two driver stages and a voltage variable attenuator (VVA). Instead of using a VGA, a variable attenuator is chosen due to its superior linearity. However, it is important to note that the input and output impedance of the VVA may vary significantly depending on the attenuation level. To prevent frequency pulling caused by impedance changes in the VVA, the two driver stages provide isolation between the mixer and the RF output. As the chip has a moderate power output of only +19 dBm, an external PA is required to boost the output power.

The ADMV4530 is equipped with a programmable four-wire serial port interface (SPI) [16] to facilitate communication with a host controller. SPI is a widely used communication method, alongside I2C [17] and UART [18], for interconnecting various chips. With the SPI functionality of the ADMV4530, the quadrature phase can be adjusted to achieve optimal sideband suppression. Furthermore, the SPI enables the nulling of LO feedthrough in IF mode. In I/Q mode, the LO feedthrough can be eliminated by applying an external DC offset to the differential baseband I/Q inputs.

3.4.4 Transceiver Example

In previous chapters, design examples for general receivers and transmitters were explored. However, the design of the SX1276 transceiver chip from Semtech [19] Corporation becomes valuable to discuss due to the crucial role of bidirectional communication in wireless devices and systems. This IC has been specifically developed for wireless communication, with a focus on IoT and low-power, long-range applications. It belongs to Semtech's SX12xx series of transceivers, designed for long-range communication utilizing LPWAN technologies like LoRaWAN [20]. The SX1276 IC is a transceiver capable of wireless data transmission and reception. It operates within sub-GHz frequency bands, typically around 868 and 915 MHz. With its robust performance, it facilitates reliable communication over several kilometers in diverse environments, including rural and urban settings.

Key features of the SX1276 IC include support for long-range communication, low power consumption, high sensitivity, and the utilization of spread spectrum modulation techniques such as chirp spread spectrum (CSS) [21]. In CSS modulation, the IC generates a chirp signal that linearly sweeps across a range of frequencies and encodes the digital data into it. By employing a spreading factor, the IC expands the encoded chirp signal over a wider bandwidth, resulting in improved signal quality, resistance to interference, and extended range. The adoption of CSS modulation in the SX1276 IC enhances its robustness, immunity to interference, and ability to facilitate long-range communication. During the transmission process, the encoded chirp signal is transmitted using the selected spreading factor, occupying a broader frequency spectrum. This expansion enhances the resistance of the signal to interference and noise. On the receiver side, the SX1276 IC listens for incoming signals within the predefined frequency range. It carries out a correlation process to detect and decode the spread spectrum chirp signals, compensating for frequency and timing variations. This compensation enables the recovery of the original encoded data, ensuring reliable data reception.

A simplified block diagram of the SX1276 is depicted in Figure 3.8. This transceiver operates in a half-duplex mode and employs a low-IF architecture. The

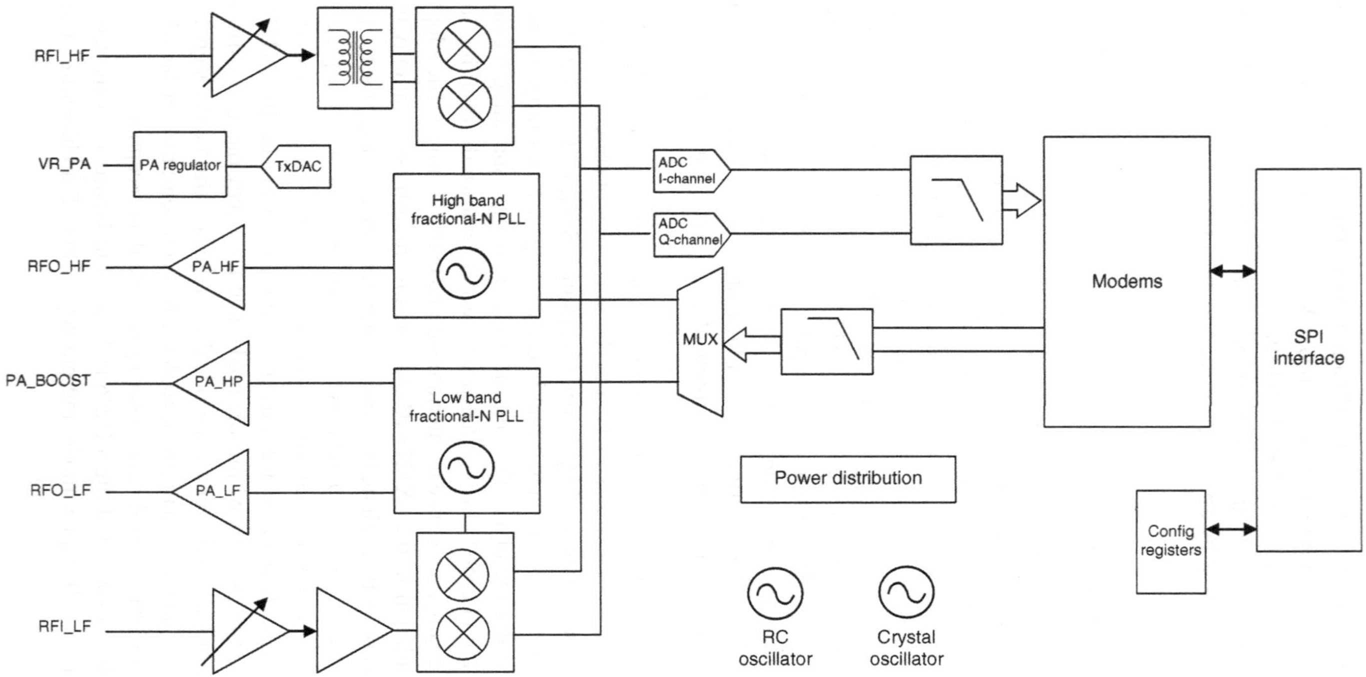

Figure 3.8 Simplified block diagram of SX1276

received RF signal first undergoes amplification by the LNA. To simplify the design process and reduce the need for external components, the LNA inputs are single-ended. After passing through the LNA inputs, the signal is converted to a differential format, enhancing its second-order linearity and rejecting harmonics. The signal is then down-converted to I&Q components at the IF by the mixer stage. Subsequent data conversion is performed by a pair of sigma-delta Analog-to-Digital Converters (ADCs) [22], and all subsequent signal processing and demodulation take place in the digital domain. Various functionalities, including automatic frequency correction (AFC), received signal strength indicator (RSSI), and AGC, are controlled by the digital state machine. Moreover, the transceiver incorporates higher-level packet and protocol functionalities through the top-level sequencer (TLS), which is exclusively available with traditional FSK and on–off keying (OOK) modulation schemes.

Fractional-N frequency synthesizers are utilized by the transceiver to generate the LO frequencies for both the receiver and transmitter. The PLLs are specifically optimized for low lock time and fast autocalibration, ensuring seamless operation without requiring user intervention. In the transmission process, FM takes place digitally within the bandwidth of the PLL. Additionally, the PLL provides an optional pre-filtering capability for the bit stream, enhancing spectral purity. Through the utilization of an SPI interface, an external host microcontroller can comprehensively configure the RF front end and digital state machine parameters of the SX1276. This enables access to the configuration registers, facilitating swift transitioning and calibration between intermediate modes of operation with the assistance of a mode auto-sequencer.

A typical application circuit utilizing the SX1276 is illustrated in Figure 3.9. This circuit necessitates only a minimal number of external components to form a complete transceiver system. An external 32 MHz crystal oscillator is required to establish the main timing reference. This oscillator serves as the reference for frequency synthesis in the PLL and acts as the clock signal for all digital processing. In order to accommodate the half-duplex operation of the transceiver, the circuit includes an RF switch. This switch is responsible for combining the transmit (TX) and receive (RX) paths into a single antenna using time division. Additionally, matching networks are integrated into the TX output and RX input to optimize power transfer and suppress out-of-band signals. These components play a crucial role in ensuring efficient transmission and reception performance.

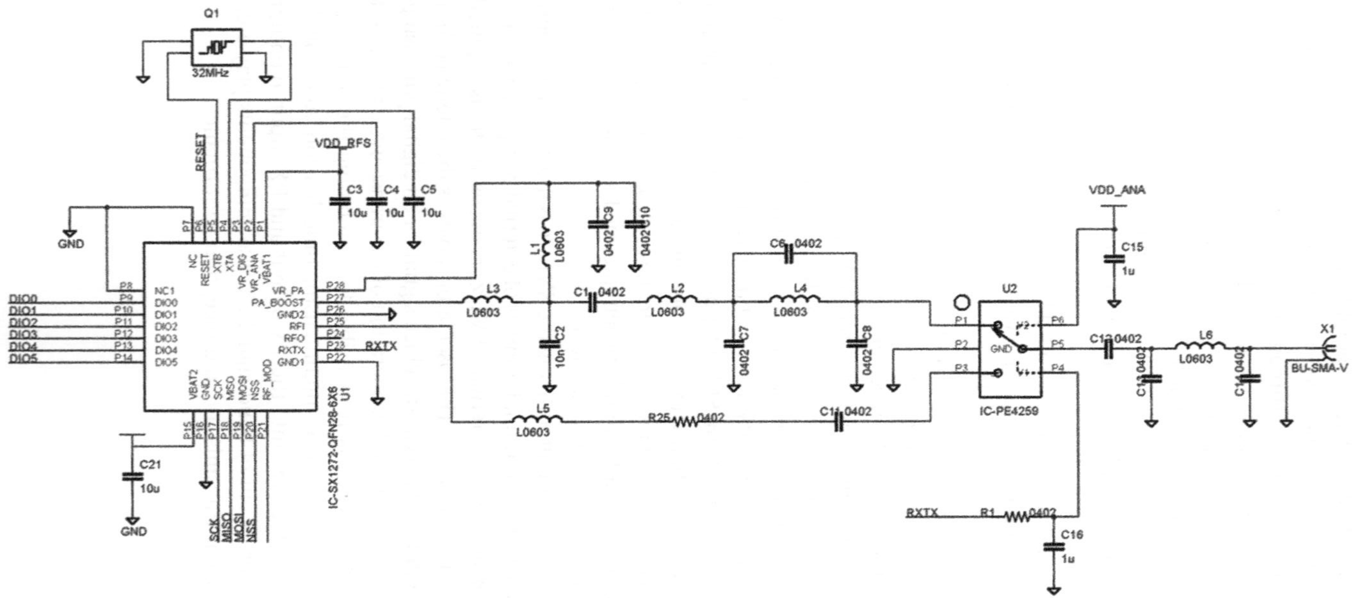

Figure 3.9 Application circuit utilizing the SX1276

References

1 Vizmuller, P. (1995). *RF Design Guide: Systems, Circuits, and Equations.* Artech House.

2 Grebennikov, A. (2011). *RF and Microwave Transmitter Design.* John Wiley & Sons Inc.

3 Baudin, P. (2015). *Wireless Transceiver Architecture - Bridging RF and Digital Communications.* John Wiley & Sons, Ltd.

4 Leeson, D.B. (1966). "A simple model of feedback oscillator noise spectrum," *Proceedings of the IEEE* 54 (2).

5 Emmerich, J. and Rudolph, H. (2023). "The trinity of inaccuracy: phase noise, jitter and short-term stability—what everyone should know about their measurement and interrelationships," *Microwave Journal.*

6 Razavi, B. (2012). *RF Microelectronics*, 2e. Pearson Education, Inc.

7 Gonzalez, G. (2003). *Microwave Transistor Amplifiers: Analysis and Design*, 2e. Prentice Hall.

8 Pozar, D.M. (2011). *Microwave Engineering*, 4e. Wiley.

9 Luzzatto, A. and Haridim, M. (2017). *Wireless Transceiver Design—Mastering the Design of Modern Wireless Equipment and Systems*, 2e. John Wiley & Sons, Ltd.

10 Rappaport, T.S. (2001). *Wireless Communications: Principles and Practice*, 2e. Prentice Hall.

11 Proakis, J.G. and Salehi, M. (2002). *Communication Systems Engineering*, 2e. Prentice Hall.

12 Skolnik, M.I. (2008). *Radar Handbook*, 3e. McGraw-Hill Professional.

13 Analog Devices. (2023). "ADMV4530D - Dual-Mode, Ka Band Upconverter with Integrated Fractional-N PLL and VCO," *ADMV4530D Datasheet.*

14 Sklar, B. (2001). *Digital Communications: Fundamentals and Applications*, 2e. Prentice Hall.

15 Haykin, S. (2005). *Communication Systems*, 4e. Wiley.

16 Mikgrusin. (2024). "SPI Basics," SparkFun Electronics Tutorial, https://learn.sparkfun.com/tutorials/serial-peripheral-interface-spi/all.

17 NXP Semiconductors (2021). *UM10204 I 2C-bus specification and user manual.* NXP Semiconductors.

18 Peňa, E. and Legaspi, M. G. (2024). "UART: a hardware communication protocol understanding Universal Asynchronous Receiver/Transmitter," https://www.analog.com/en/resources/analog-dialogue/articles/uart-a-hardware-communication-protocol.html.

19 Semtech Corporation. (2020). "SX1276/77/78/79 - 137 MHz to 1020 MHz low power long range transceiver," *Datasheet.*

20 The Things Network, "Everything you need to know about LoRaWAN in 60 minutes," *Technical Tutorial*, https://www.thethingsnetwork.org/docs/lorawan/.

21 Simon, M.K. and Omura, J.K. (1994). *Spread Spectrum Communications Handbook*. McGraw-Hill.

22 Analog Devices (2005). *The Data Conversion Handbook*. Analog Devices https://www.analog.com/en/education/education-library/data-conversion-handbook.html.

4

Software-Defined Radio

4.1 Introduction

Wireless communication has evolved significantly, starting with radio transmissions, and progressing to cell phones and modern smartphones. Software-defined radio (SDR) offers advantages such as flexibility, upgradability, and reconfigurability. SDR relies on digital signal processing (DSP) algorithms, reducing the need for analog hardware. It improves reliability, enhances spectrum utilization, and finds applications in various domains. An overview of SDR is provided, highlighting its advantages over conventional radio systems. The underlying concepts and architecture of SDR are examined, including direct synthesis for compact implementation. Key building blocks, such as the antenna, RF front-end, and analog-to-digital and digital-to-analog converters (ADCs and DACs), are explored.

A deeper look is taken into the individual building blocks, discussing their functionalities and design considerations. The utilization of software-defined antennas (SDAs) to leverage SDR capabilities is discussed, along with the importance of the RF front-end in signal transmission and reception. The vital process of analog-to-digital and digital-to-analog conversion is explored, bridging the gap between the analog and digital domains. The digital front-end (DFE) plays a pivotal role in SDR systems as it performs functions such as Sample Rate Conversion (SRC) for synchronization, Channelization for dividing bandwidth, Digital Up Conversion (DUC) for transmission, and Digital Down Conversion (DDC) for reception. Signal processing, the core of SDR systems, is discussed, enabling the manipulation, analysis, and enhancement of signals in both the transmitter and receiver sides. It encompasses functions such as modulation, demodulation, channel equalization, signal detection, synchronization, and error correction coding.

A design example for an indoor Global Positioning System (GPS) system is presented to showcase the practical implementation of an SDR system, illustrating how the building blocks and concepts come together.

Introduction to Wireless System Design: From Circuits to Web-based Applications, First Edition.
Henry Lau, Ludy Liu, and Keith Chan.

4.2 A New Radio

Since its beginning in the late 19th century [1–3], wireless communication has undergone significant advancements. It started with radio transmissions and the introduction of analog voice and television communications. The field then progressed with the development of wired and wireless computer networks such as ARPANET [4] and ALOHANET [5]. While early cell phones [6] were initially bulky, they revolutionized wireless voice communications.

In the modern era, smartphones have transformed into portable computers that offer high-speed access to cellular networks and the Internet. The introduction of Wireless Local Area Networks in 1985, regulated by the FCC, played a crucial role in the development of protocols like Wi-Fi and Bluetooth, which became essential for corporate networks [7]. However, many wireless devices face a significant limitation due to their reliance on hardware-based radios and protocols. This lack of flexibility presents challenges when it comes to reprogramming or reconfiguring these devices, particularly concerning radio functions. Rectifying errors in hardware, firmware, or software can prove to be difficult, and sometimes even impossible. Furthermore, these devices are typically confined to the functionality provided by their hardware components, making it challenging to support other wireless protocols without significant modifications.

SDR offers numerous advantages [8–10], such as design flexibility, upgradability, and reconfigurability. Its software-defined nature enables seamless software updates and enhancements, reducing costs and ensuring compatibility with different wireless standards and waveforms. SDR devices can be easily reconfigured to operate on various frequencies and modulation schemes, enhancing their adaptability to evolving communication requirements.

Compared to traditional hardware-based receivers, SDR technology improves reliability. Hardware components can exhibit performance variations due to factors like temperature and aging, whereas SDR systems provide consistent performance over time since their functionalities are implemented in software. Additionally, SDR has the potential to enhance spectrum utilization and efficiency. Through advanced signal processing techniques, SDR systems can dynamically allocate and utilize available frequency bands, minimizing interference and maximizing spectrum usage.

SDR primarily relies on software-based DSP algorithms to handle the majority of physical layer functions. Instead of utilizing traditional analog components for mixing, amplification, filtering, and modulation/demodulation, SDR employs digital implementations, reducing the dependence on discrete analog hardware. In an optimal configuration, an SDR setup minimizes hardware requirements at the RF front-end, typically consisting only of an antenna and a high-speed sampler capable of capturing and digitizing wideband radio signals.

Transitioning from discrete analog components to an embedded platform offers several advantages for wireless communication system designers [9]:

a) Design flexibility: The availability of fast and accurate digital-to-analog and analog-to-digital converters allows for fine-tuning of traditional hardware circuitry, such as filtering and analog modulators/demodulators.
b) Reliability: Hardware components can exhibit variations in performance due to factors like temperature, manufacturing variations, and aging, which can lead to performance degradation in communication systems. Implementing these functions in an SDR platform helps maintain consistent system performance over time.
c) Upgradability: Upgrading and improving a wireless system composed of discrete components often requires replacing hardware, resulting in additional costs. With an SDR, system upgrades can be performed through software updates, providing a more cost-effective solution.
d) Reusability: Once a physical layer (PHY) is designed and tested in software, it can be deployed on different hardware platforms, promoting reusability and reducing development time.
e) Reconfigurability: Current SDR implementations offer hardware reconfigurability, allowing support for multiple wireless standards and waveforms. This flexibility enables adaptation to different communication requirements.
f) Lower cost: Building a communication system from scratch requires acquiring numerous hardware components, which can be expensive. SDR devices, on the other hand, tend to be more cost-effective due to their software-based approach

SDR finds applications in diverse domains, including telecommunications, military communications, public safety, amateur radio, satellite communication, cognitive radio, and research and development. Its flexibility, adaptability, and high performance make it well-suited for a wide range of wireless communication scenarios.

4.3 Concepts and Architecture

4.3.1 Direct Synthesis

An SDR refers to a system that primarily relies on DSP algorithms implemented in an embedded system using specific software to handle most of the physical layer functions. In an SDR, the traditional analog stages involved in establishing a wireless radio link, such as mixing, amplification, filtering, and modulation/demodulation, are replaced by digital implementations [11]. This implies that the majority

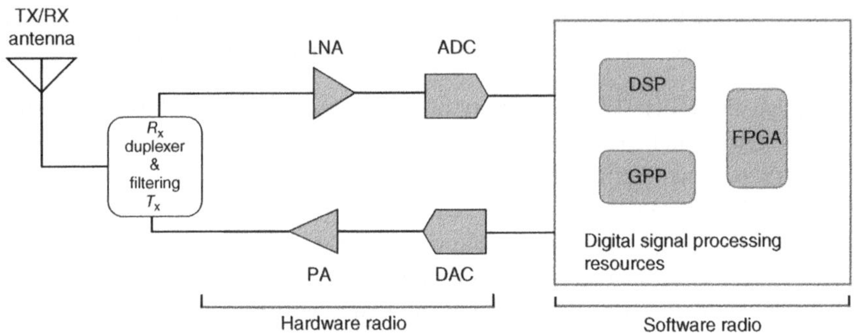

Figure 4.1 An ideal SDR transceiver block diagram

of physical layer functionalities are performed digitally rather than using discrete analog hardware components.

As depicted in Figure 4.1, an ideal SDR [12] comprises minimal hardware components at the RF front-end, primarily consisting of an antenna and a high-speed sampler capable of capturing and digitizing wideband radio signals. However, in certain scenarios, amplifiers may be utilized before the DAC/ADC stages to achieve extended coverage distances. For instance, at the receiver, a Low-Noise Amplifier (LNA) is employed to reduce quantization noise during conversion and maximize the Signal-to-Noise Ratio (SNR) in the digital domain after sampling. Once the signal is converted to binary form, it undergoes processing by dedicated computational units within the embedded system. These computational units facilitate the implementation of crucial methods for demodulation, synchronization, and decoding, which are necessary for recovering transmitted information, particularly from advanced modulation techniques.

The primary function of the transmitter, as shown in Figure 4.1, is to convert the digital samples computed in the baseband subsystem into an RF waveform at the desired carrier frequency [13]. This RF waveform is subsequently transformed into an electromagnetic wave by the antenna. The simplest form of the transmitter front-end is based on direct RF synthesis, as depicted in Figure 4.2.

In this configuration, the DSP subsection incorporates a digital frequency synthesizer with an effective clock rate that is at least twice the desired carrier frequency, F_c. The DAC sample rate, F_s, must be greater than $2 * F_c$. The spectrum of the generated signal is illustrated in Figure 4.2. Aliases of the desired signal occur at multiples of the sampling frequency and must be filtered out before reaching the power amplifier (PA). It is important to note that as F_c approaches $F_s/_2$, the separation between the desired signal and the alias diminishes, necessitating a sharper filter. Practical limitations on filter sharpness restrict the maximum supported F_c to a value well below $F_s/2$.

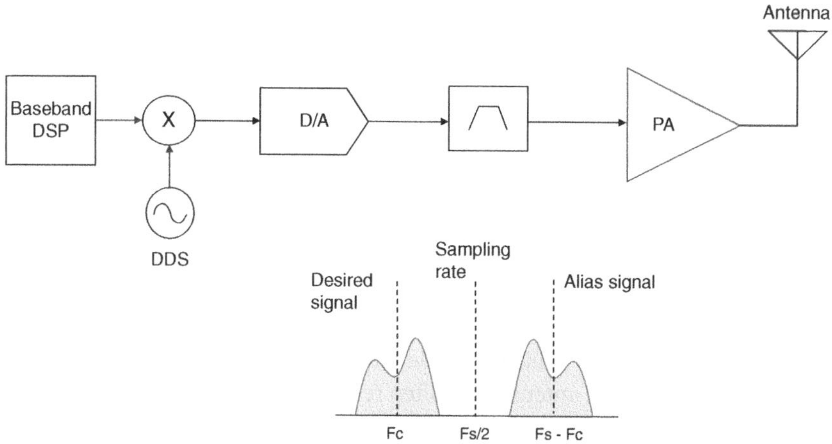

Figure 4.2 Direct RF synthesis on transmitters

Direct RF synthesis offers several advantages. Firstly, it reduces the number of components, resulting in a more compact implementation and potentially lower cost. Additionally, it minimizes sources of noise and distortion. Another advantage is its ability to achieve very agile frequency synthesis, allowing for rapid frequency changes without significant delays. This feature is particularly beneficial for cognitive radios and wideband frequency-hopped systems. However, despite these advantages, direct RF synthesis is not commonly used in modern radios due to certain disadvantages. One major drawback is the high sample rate requirements for the DAC. For example, generating a 2.4 GHz carrier frequency necessitates a minimum sample rate of 5 GHz. While DACs with such sample rates do exist, they are expensive and consume a significant amount of power. Moreover, generating even higher carrier frequencies directly is not feasible due to the absence of suitable DACs.

Another disadvantage is that high-speed DACs offer lower accuracy and dynamic range (fewer bits) compared to their lower-speed counterparts. Modern waveforms like OFDM demand a high dynamic range, and even more dynamic range is necessary if the transmitter supports multiple simultaneous users with varying power levels. These limitations make direct RF synthesis less practical in many radio applications.

On the receiver side, the received signal picked up by the antenna is typically amplified by an LNA, then sampled and digitized into the digital domain directly for further DSP. This direct sampling approach has the same advantage as direct RF synthesis on the transmitter—fewer components and easier implementation. However, the major drawback is the high sample rate requirements for the ADC.

4.3.2 Building Blocks

As mentioned in the previous section, digital-only implementation in SDR design has not come yet. Most SDR systems are still using a more classical approach as shown in Figure 4.3. A typical SDR transceiver consists of four building blocks, namely, the Antenna, Analog RF Front-end, DFE, and Signal Processing [14].

Figure 4.3(a) and (b) provide a high-level overview of a typical SDR transceiver, which comprises the main building blocks as: signal processing, DFE, analog RF front-end, and an antenna.

4.3.2.1 Antenna

SDR platforms commonly utilize multiple antennas to cover a wide range of frequency bands [15]. These antennas are often referred to as "intelligent" or "smart" due to their capability to select frequency bands and adapt through mobile tracking or interference cancellation [16, 17]. In the context of SDRs, antennas must meet specific requirements, such as self-adaptation (flexibility to tune to multiple bands), self-alignment (beamforming capability), and self-healing (interference

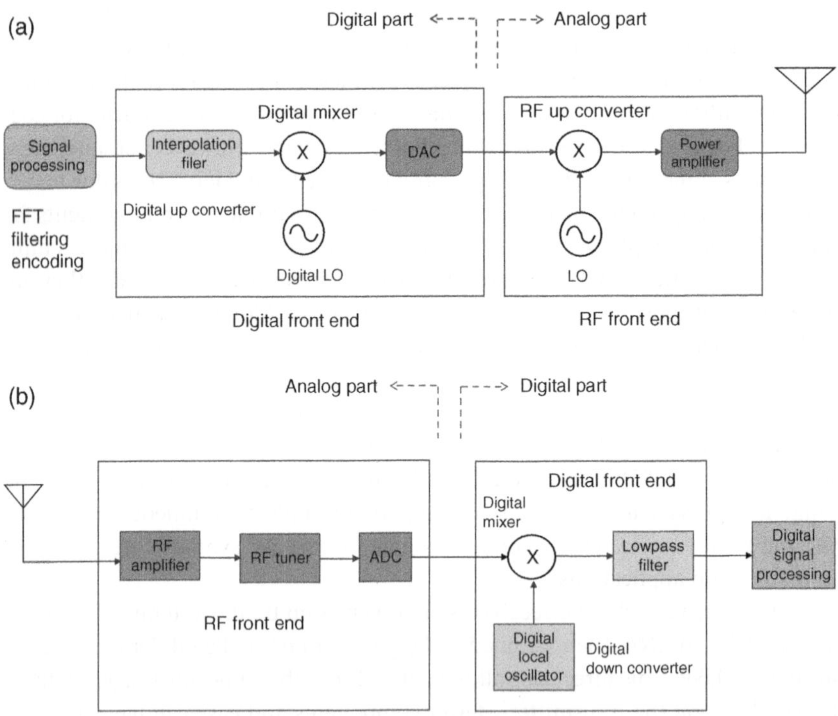

Figure 4.3 (a) Classical SDR transmitter and (b) Classical SDR receiver

rejection). SDA technology leverages SDR capabilities to enable dynamic control and adaptation of antenna operating characteristics. Unlike traditional antennas with fixed properties, SDAs utilize the flexibility and adaptability provided by SDR to achieve the desired performance.

An SDA system typically consists of a DFE that includes an SDR module and an array or elements of reconfigurable antennas. The SDR module processes incoming and outgoing signals, enabling real-time control, and adjustment of antenna parameters. The reconfigurable antenna elements allow for changes in frequency, radiation pattern, beam direction, and polarization. The SDR module enables the application of advanced signal processing techniques such as beamforming and software-defined networking (SDN). Beamforming dynamically steers the antenna's beam towards specific targets, improving signal reception, and transmission in desired directions. SDN enables the antenna to adapt its characteristics to optimize performance based on network requirements.

SDA offers several advantages over traditional antennas. It provides increased flexibility and agility, allowing for rapid reconfiguration to adapt to changing communication needs. It optimizes spectrum utilization by dynamically allocating frequencies and adjusting radiation patterns. It also improves system performance through adaptive interference mitigation and enhanced signal quality. SDA technology finds applications in various wireless communication systems, radar systems, satellite communication, and Internet of Things (IoT) networks. It has the potential to revolutionize future wireless systems by providing intelligent and adaptable antenna solutions.

4.3.2.2 RF Front-end

The RF Front-end is a circuitry that handles the transmission and reception of signals at various operating frequencies. It also converts the signal to/from the Intermediate Frequency (IF). The operation of the RF Front-end is divided into two paths, depending on the signal direction (Tx or Rx mode). In the transmission path, digital samples are converted into an analog signal by a DAC. The analog signal is then mixed with a predetermined RF frequency, modulated, and transmitted. The transmitter architectures are the same as those mentioned in Chapter 3, such as the direct-conversion and heterodyne approaches.

In the receiving path, the antenna captures the RF signal, which is then connected to the RF Front-end through a matching circuitry to optimize signal power transfer. The signal passes through an LNA located near the antenna to amplify weak signals and minimize noise. The amplified signal, combined with a signal from the Local Oscillator (LO), is fed into a mixer to down-convert it to the IF [18]. Receiver architectures closely mirror the transmitter architectures (direct RF sampling, zero-IF downconversion, IF downconversion, and super-heterodyne downconversion).

4.3.2.3 Analog-to-Digital and Digital-to-Analog Conversion

In SDR applications, the DAC is responsible for generating the analog signal from the digital samples, while the ADC plays a crucial role in converting continuous-time signals into discrete-time, binary-coded signals on the receiver side. When considering the requirements for ADC and DAC in SDR applications, several factors should be taken into account [19, 20]:

a) Sampling rate: SDR applications often demand high sampling rates to effectively capture and process wideband signals. The ADC should be capable of sampling at a rate that satisfies the Nyquist criterion, which states that the sampling rate should be at least twice the highest frequency component of the signal being sampled. Similarly, the DAC should support a sufficiently high sampling rate to accurately reconstruct the analog signal.
b) Resolution: The resolution of the ADC and DAC determines the level of detail in the captured and reconstructed signals. Higher bit resolutions provide greater precision, but they may come at the cost of increased complexity and power consumption. The required resolution depends on the specific application's dynamic range and SNR requirements.
c) Linearity and dynamic range: Linearity refers to the ability of the ADC or DAC to accurately represent the input or output signal over a wide range of amplitudes. A high-quality ADC/DAC should exhibit low distortion, minimal harmonic and intermodulation distortion, and a wide dynamic range to faithfully capture and reproduce the analog signals without introducing significant nonlinearities.
d) Noise performance: ADCs and DACs introduce some level of noise into the signal. In SDR applications, it's important to minimize noise to maintain good SNR and minimize signal degradation. ADCs and DACs characterized by low noise figures, low spurious signals, and favorable signal-to-noise plus distortion ratios (SINAD) are preferred.
e) Bandwidth: SDR applications often deal with wideband signals, so the ADC and DAC should have sufficient bandwidth to handle the desired frequency range. The analog front-end components, such as anti-aliasing filters for the ADC and reconstruction filters for the DAC, should also be designed to match the required bandwidth.
f) Power consumption: SDR applications may involve portable or battery-powered devices where power consumption is a critical factor. Consider ADCs and DACs with low power consumption while still meeting the performance requirements of the application.

When choosing ADC and DAC models for SDR applications, the selection depends on the specific requirements of the application. Within the SDR community, there are several ADC and DAC models that have become popular choices.

One widely used ADC model is the Analog Devices AD9361 [21]. It is a highly integrated RF transceiver designed specifically for SDR applications. The AD9361 features dual ADCs and DACs, supports a wide frequency range, and offers high-performance signal processing capabilities. Another notable ADC model is the Texas Instruments ADS58C20 [22]. This dual-channel ADC has 11-bit resolution and can sample at a rate of 800 MSPS. It delivers excellent dynamic performance and is well-suited for high-speed SDR applications. In terms of DAC models, the Analog Devices AD9744 [23] is often preferred. It is a high-speed, low-power 14-bit DAC that supports sampling rates up to 210 MSPS. The AD9744 delivers excellent dynamic performance and is commonly used in SDR applications. For demanding SDR applications, the Texas Instruments DAC38RF82 [24] is a notable choice. This dual-channel DAC offers 16-bit resolution and a sampling rate of 9 GSPS. It provides high-speed and high-resolution performance, making it suitable for advanced SDR applications.

4.3.2.4 Digital Front-end

The DFE in SDR systems plays a vital role in signal processing and communication. It performs several important functions [19] that are essential for preparing the signals for further digital processing and conversion. One key function of the DFE is SRC. SRC involves converting the sampling rate of the incoming signal to a different rate. This is necessary to ensure synchronization between different components or communication parties in the SDR system. By adjusting the sampling rate, SRC enables the signals to match the requirements of subsequent processing stages or interface with other devices.

Another important function of the DFE is Channelization. Channelization refers to dividing the available bandwidth into smaller frequency channels. In the DFE, channelization involves extracting specific frequency bands of interest from the received signal. This can include tasks such as filtering, demultiplexing, and separating different channels. Channelization is particularly relevant in multichannel SDR systems where multiple signals need to be processed simultaneously.

On the transmitter side of an SDR system, the DFE performs DUC. DUC translates the baseband signal to an IF or RF frequency for transmission. This is achieved through digital mixing, interpolation, and filtering techniques. By applying DUC, the DFE prepares the signal for transmission over the air or through other communication channels. Conversely, on the receiver side, the DFE performs DDC. DDC involves converting the received RF signal to a lower IF or baseband signal for further processing. The DDC utilizes digital mixing, decimation, and filtering techniques to downconvert the signal to a lower frequency range suitable for subsequent digital processing stages.

The DFE also includes functions for Signal Conditioning. Signal conditioning tasks are aimed at preparing the signals for optimal processing and performance. This can involve operations such as amplification, gain control, filtering, and noise suppression. By enhancing the quality and reliability of the signals, signal conditioning helps ensure accurate and reliable processing. Overall, the DFE in SDR systems plays a critical role in bridging the gap between analog and digital domains. Its functions enable efficient signal processing and enhance the performance of SDR systems in various applications.

4.3.2.5 Signal Processing

Signal processing plays a critical role in SDR systems. It involves a range of functions that enable the manipulation, analysis, and enhancement of signals in both the transmitter and receiver sides. These functions are essential for achieving reliable and efficient communication. One of the key functions in SDR signal processing is modulation and demodulation. Modulation involves encoding digital information onto a carrier wave for transmission, while demodulation is the process of extracting the original information from the received modulated signal. Various modulation schemes, such as AM, FM, PM, QAM, or PSK, are used to enable signal transmission and reception.

Another important function is channel equalization, which compensates for impairments introduced by the communication channel. Signal processing techniques, including equalization algorithms, are employed to counteract distortions and attenuations, improving the overall quality of the received signal. This enhances the reliability of communication. Signal detection and synchronization are crucial tasks in SDR systems. Signal processing algorithms are utilized to detect and synchronize with the received signal. This involves estimating the carrier frequency and phase, recovering the symbol timing, and achieving frame synchronization. Accurate signal detection and synchronization are vital for extracting the transmitted information reliably and preparing it for further processing. Error correction coding and decoding are employed to mitigate errors introduced by the communication channel. These techniques enhance the system's ability to detect and correct errors at the receiver side. Signal processing algorithms are used to encode the data at the transmitter using forward error correction (FEC) codes and decode it at the receiver, improving the overall reliability of the communication system.

The block responsible for signal processing is commonly referred to as the baseband processing block. It holds a central position in discussions about SDRs due to its significant presence in the system's digital domain implementation. Its implementation relies on specialized hardware circuitry designed specifically for efficient signal processing. Various hardware platforms, including Digital Signal Processors, Field-Programmable Gate Arrays (FPGAs) [16], Application-Specific

Integrated Circuits (ASICs) [25], General-Purpose Processors (GPPs) [26], and Graphics Processing Units (GPUs) [27], are utilized for this purpose. These hardware components play a crucial role in enabling high-performance signal processing capabilities. Consequently, SDR systems can effectively manipulate and analyze signals within the digital domain, resulting in improved functionality and performance.

ASICs are custom-designed integrated circuits optimized for specific applications. They excel at performing dedicated functions efficiently, making them ideal for SDR tasks that require high performance and low power consumption. However, ASICs are expensive to design and manufacture, which limits their use to high-volume applications where cost and power efficiency are critical. On the other hand, FPGAs are programmable logic devices that offer flexibility in hardware design. They can be configured and reconfigured to implement various digital circuits, making them well-suited for SDR systems. By programming an FPGA, one can define the functionality and signal processing chain of an SDR system. FPGAs strike a balance between performance, flexibility, and cost, making them a popular choice for SDR implementations.

DSPs are specialized microprocessors designed specifically for efficient execution of DSP tasks. They excel at performing mathematical operations and algorithms on digital signals in real time. In SDR systems, DSPs are often responsible for tasks such as filtering, demodulation, and decoding. They offer a good balance between performance and power efficiency and can be programmed using specialized software development tools. GPPs, or general-purpose processors, refer to processors like CPUs found in typical computer systems. While GPPs are not specifically designed for real-time signal processing, they can be utilized in SDR systems for non-real-time tasks such as control, user interface, and networking. However, their performance may be limited for computationally intensive signal processing operations.

Lastly, GPUs are specialized processors primarily designed for accelerating graphics rendering, but their parallel processing capabilities make them well-suited for certain signal processing tasks. They excel at performing large-scale parallel computations, making them useful in SDR applications that involve computationally intensive tasks such as fast Fourier transforms (FFTs), channel estimation, and beamforming.

4.4 Design Example

Also introduced in Chapter 1, the GPS has gained immense popularity [28] due to its extensive coverage, reliability, and accuracy. It comprises a constellation of satellites orbiting the Earth, along with ground control stations and GPS receivers that users utilize to determine their precise location, navigate routes, and obtain precise timing information. One of the key factors contributing to the popularity of GPS is its broad range of applications. GPS technology is integrated into various devices that individuals use in their daily lives. Smartphones, smartwatches, and portable navigation devices (PNDs) are some examples of devices equipped with GPS receivers. This integration allows users to navigate, access location-based services, and receive real-time traffic updates with ease.

The transportation and logistics industry heavily relies on GPS for efficient operations. Fleet management systems utilize GPS to track vehicles, optimize routes, monitor driver behavior, and improve overall efficiency. Additionally, GPS plays a vital role in aviation, enabling accurate navigation, approach procedures, and aircraft tracking. Systems like the Wide Area Augmentation System (WAAS) enhance the precision and reliability of GPS-based navigation in aviation. Marine navigation also benefits greatly from GPS. It provides accurate positioning information for route planning, collision avoidance, and search and rescue operations at sea. Marine GPS systems offer essential data to ships, boats, and other vessels, ensuring safe and efficient navigation.

While GPS is the most popular navigation system, it is important to note that other Global Navigation Satellite Systems (GNSS) also exist. Examples include GLONASS (Russia), Galileo (European Union), BeiDou (China), and NavIC (India). Many modern GPS receivers are designed to work with multiple GNSS systems, leveraging signals from different constellations to enhance accuracy and reliability.

Clear and reliable reception of GPS/GNSS signals is contingent upon having an unobstructed view of the sky, ensuring a strong signal is received. However, when navigating through a tunnel, this view is impeded, rendering the signals inaccessible. In such scenarios, mapping software compensates for the temporary signal loss by assuming the vehicle is safely progressing within the tunnel. Nonetheless, this compensation may only be effective for a limited duration, as deviations from the actual position accumulate over time. In extended tunnels, the accumulated positional error can become significant, posing challenges for both accurate compensation and route planning immediately upon exiting the tunnel.

With the continuous advancement of autonomous driving technology, the reliable tracking of GPS signals within tunnels has become increasingly crucial. This is especially true in densely populated cities where tunnels are prevalent throughout the road network. Accurate GPS tracking inside tunnels is essential for the seamless integration of autonomous vehicles and efficient navigation in urban environments characterized by numerous driving tunnels. Therefore, having GPS signals available in indoor environments such as tunnels is highly desirable.

To bring GPS signals into indoor environments, GPS repeaters are commonly used. Traditional repeaters involve installing an outdoor GPS antenna, along with an amplifier, and using a coaxial cable to transmit the signal to an indoor area equipped with a radiating antenna or leaky coaxial cable. However, this approach presents some challenges. The first challenge is that GPS receivers receiving the repeated signal may mistakenly interpret their location as being at the outdoor antenna, rather than their actual indoor position. This can make it difficult to accurately track individuals, particularly if they are underground or far away from the outdoor antenna. The second challenge arises when a leaky cable is used. In such cases, the signal strength is strongest near the outdoor antenna amplifier, potentially causing GPS disruptions for users in the immediate vicinity.

To overcome these limitations, an Intelligent Repeater System [28, 29] provides a solution. In this system (as shown in Figure 4.4), an outdoor antenna is installed in a location with an unobstructed view of the sky.

Instead of directly connecting the antenna to an amplifier and antenna system, the Intelligent Repeater System incorporates a data collection and synchronization unit. This unit collects information about live sky signals and generates precise 10 MHz and 1PPS signals. These signals can be used by a GNSS simulator to recreate live sky signals. However, unlike traditional repeater systems that generate

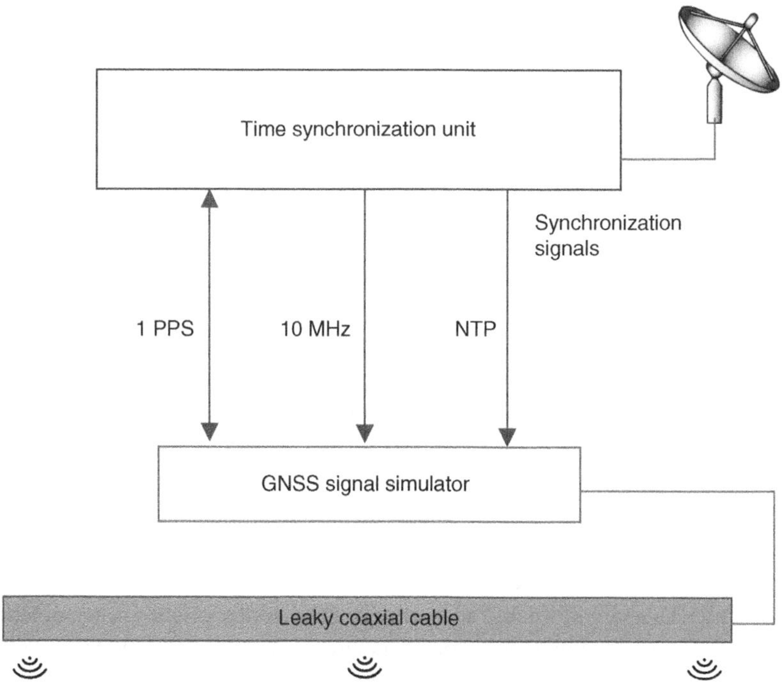

Figure 4.4 Single zone Intelligent Repeater System

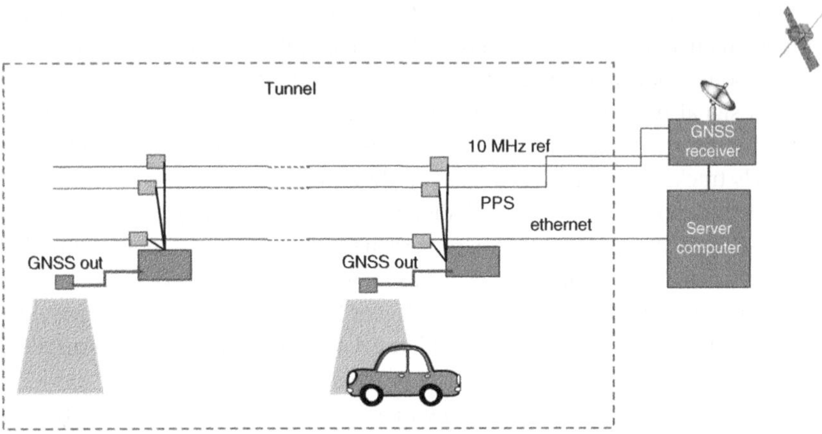

Figure 4.5 Zone-based Intelligent Repeater System

signals corresponding to the outdoor antenna location, the Intelligent Repeater System allows the generated signals to be programmed to simulate positions anywhere in the world within the indoor zone. By utilizing the Intelligent Repeater System, personnel or equipment can be accurately located within a specific indoor zone as shown in Figure 4.5.

This capability significantly reduces critical response times, enhances safety, and provides more accurate information compared to traditional repeater systems. The Intelligent Repeater System effectively addresses the limitations of conventional repeaters by enabling indoor positioning without signal mislocation and local disruptions, thereby improving the overall effectiveness of indoor tracking systems.

The Intelligent Repeater System comprises a satellite receiver, which is responsible for collecting GNSS data and synchronization signals, and multiple GNSS Simulators for generating GNSS RF signals. When it comes to selecting a suitable device for the GNSS simulator, an SDR device is highly recommended [30, 31]. With its software configurability, an SDR device can effectively generate the necessary GPS RF signals within the L-band frequency range, typically ranging from 1.2 to 1.5 GHz. This generation is based on the GNSS data collected by the satellite receiver and the simulated location within each specific zone. Various SDR products are available for GPS simulation, including HacRF One, USRP, and BladeRF [32–34]. The HacRF One is used for the SDR demonstration.

The HackRF One, an affordable and versatile SDR device, offers a range of features that make it a valuable tool. One of its key attributes is its wide frequency range, spanning from 1 MHz to 6 GHz. This extensive coverage enables the HackRF One to receive and transmit signals across a variety of frequency bands, making it suitable for diverse applications. An outstanding characteristic of the

HackRF One is its software configurability, allowing users to fully program and customize it using SDR tools like GNU Radio. This flexibility empowers users to adapt the HackRF One to their specific requirements and tailor it for different use cases.

The HackRF One supports half-duplex operation, which means it can either transmit or receive signals at a given time. While bidirectional communication is possible, it cannot occur simultaneously. The device boasts a wide dynamic range, enabling it to handle signals of varying strengths with excellent sensitivity and minimal distortion. Connectivity is facilitated through USB, making it easy to connect the HackRF One to a computer or other devices. This USB connectivity simplifies integration with different systems, providing a straightforward and convenient setup.

Figure 4.6 shows the block diagram of the HackRF One. It incorporates several major components that play essential roles in its design and functionality. One of these components is the MAX2837 transceiver [35], operating within the frequency range of 2.3–2.7 GHz. The transceiver acts as the interface for transmitting and receiving signals within this frequency band, making it a crucial element in the HackRF One. Another significant component is the MAX5864 ADC/DAC [36]. This component enables the conversion of analog signals to digital and vice versa. By facilitating this conversion, the ADC/DAC allows the HackRF One to process and manipulate signals in the digital domain, providing greater flexibility in signal processing.

The Si5351 clock generator [37] is another integral part of the HackRF One. This component generates precise clock signals that synchronize the operations of

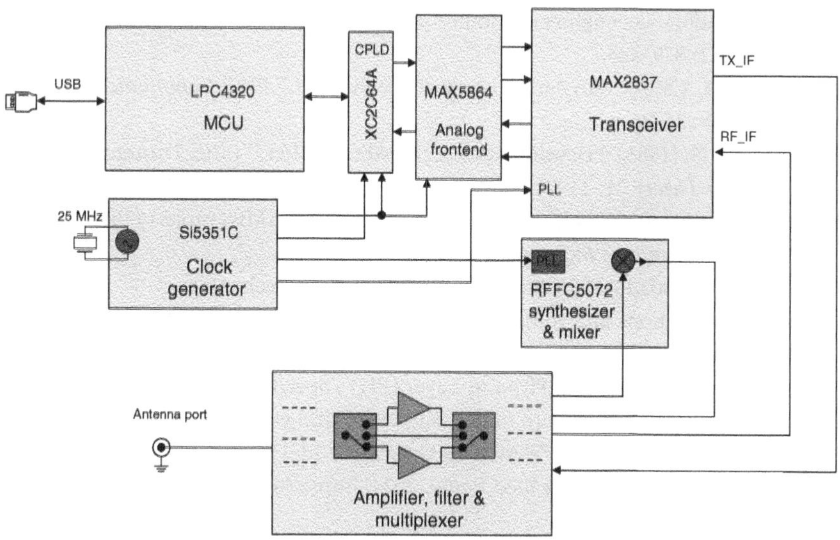

Figure 4.6 Simplified block diagram of HackRF One

various components within the device. Accurate timing and synchronization are crucial for proper functioning and coordination of the HackRF One's internal processes. The HackRF One utilizes a CoolRunner-II CPLD (Complex Programmable Logic Device) [38], which is a programmable logic device. This CPLD enables the implementation of custom logic and functionality within the device. By programming the CPLD, users can tailor the HackRF One to perform specific tasks and functions based on their requirements.

At the core of the HackRF One is the LPC43xx microcontroller [39], specifically the ARM Cortex-M4 variant. This microcontroller serves as the main processing unit, executing firmware and software instructions that control the device's operations. It handles tasks such as data processing, communication with external systems, and overall device management. Lastly, the HackRF One incorporates the RFFC5072 mixer/synthesizer [40], which combines the functions of a mixer and a synthesizer. This component is responsible for generating and manipulating RF signals within the device, allowing it to transmit and receive signals across different frequencies effectively.

References

1 Green, J.J. (1899). "The apparatus for wireless telegraphy," *American Electrician* 344–346.
2 Barnouw, E. (1966). *A Tower in Babel (A History of Broadcasting in the United States to 1933).* Oxford University Press.
3 Bondyopadhyay, P.K. and Marconi, G. (1995). "The father of long distance radio communication—an engineer's tribute," *Proceedings of the 25th European Microwave Conference* 2: 879–885.
4 Kleinrock, L. (2010). "An early history of the internet," *IEEE Communications Magazine* 48: 26–36.
5 Abramson, N. (1985). "Development of the ALOHANET," *IEEE Transactions on Information Theory* 31: 119–123.
6 Frenkiel, R. (2010). "Creating cellular: a history of the AMPS project (1971–1983)," *IEEE Communications Magazine* 14–24.
7 IEEE. (1997). "802.11 IEEE Standard for Information Technology. Telecommunications and Information Exchange Between Systems, Local and Metropolitan Area Networks. Specific Requirements. Part 11: Wireless LAN Medium Access Control (MAC) and Physical Layer (PHY) Specification".
8 Mitola, J. III (1992). "Software radios. Survey, critical evaluation and future directions," *IEEE National Telesystems Conference* 13–15.
9 Radixon Group, *Software-Defined Radio*, 2024, https://www.winradio.com/home/facts.htm

10 Machado-Fernández, J.R. (2015). "Software defined radio: basic principles and applications," *Revista Facultad de Ingeniería* 24 (38).

11 Duarte, L. et al. (2019). "A software-defined radio for future wireless communication systems at 60 GHz," *Electronics.*

12 Bronckers, S., Ro'ch, A., and Smolders, B. (2017). Wireless receiver architectures towards 5G: where are we? *IEEE Circuits System Magazine* 6–16.

13 Grayver, E. (2013). *Implementing Software Defined Radio*. Springer.

14 Akeela, R. and Dezfouli, B. (2018). "Software-defined radios: architecture, state-of-the-art, and challenges," *Computer Communications*, Elsevier 128: 106–125.

15 Rohde, U.L. and Bucher, T.T.N. (1988). *Communications Receivers: Principles and Design*. McGraw-Hill.

16 Haghighat, A. (2002). "A review on essentials and technical challenges of software defined radio," *MILCOM 2002 Proceedings* 377–382.

17 Rouphael, T.J. (2009). *RF And Digital Signal Processing for Software-Defined Radio: A Multi-Standard Multi-Mode Approach*. Newnes.

18 Carr, J.J. (2001). *The Technician's Radio Receiver Handbook: Wireless and Telecommunication Technology*. Newnes.

19 Hentschel, T., Henker, M., and Fettweis, G. (1999). "The digital front-end of software radio terminals," *IEEE Personal Communications* 6 (4): 40–46.

20 Walden, R. (1999). "Analog-to-digital converter survey and analysis," *IEEE Journal on Selected Areas in Communications* 17 (4): 539–550.

21 Analog Devices. (2016). "RF Agile Transceiver AD9361," Product Datasheet.

22 Texas Instruments Incorporated. (2022). "Dual IF Receivers with SNRBoost3G+ Signal Processing ADS58C20 and ADS58C23," Features Datasheet.

23 Analog Devices. (2013). "AD9744 14-Bit, 210 MSPS TxDAC D/A Converter," Product Datasheet.

24 Texas Instruments Incorporated. (2020). "DAC38RF8x Dual-Channel, Differential-Output, 14-Bit, 9-GSPS, RF-Sampling DAC With JESD204B Interface, On-Chip PLL and Wide-Band Interpolation," Datasheet.

25 Barr, K. (2006). *ASIC Design in the Silicon Sandbox: A Complete Guide to Building Mixed-Signal Integrated Circuits*, 1e. McGraw Hill.

26 Patterson, D.A. and Hennessy, J.L. (2020). *Computer Organization and Design: The Hardware/Software Interface*, 6e. Morgan Kaufmann.

27 Hwu, W.W. and Kirk, D.B. (2022). *Izzat El Hajj, Programming Massively Parallel Processors: A Hands-on Approach, 4e. Morgan Kaufmann.*

28 Perdue, L., Sasaki, H. and Fischer, J., "Zone Based Indoor Location Using GNSS Simulators," *International Symposium on GNSS 2015, Kyoto, Japan*, pp. 16–19.

29 Stepglobal, "A GNSS Simulation System for Real Time Tracking within Tunnels," updated to 2023, https://stepglobal.com/a-gnss-simulation-system-for-real-time-tracking-within-tunnels/

30 GPS World, "The Future of GNSS Simulation is Software-defined," last modified February 29, 2024, https://www.gpsworld.com/sponsoredcontent/the-future-of-gnss-simulation-is-software-defined/

31 Software-Defined GNSS Simulator (2015). *A Step Forward*, 54–60. Skydel Solutions, InsideGNSS.

32 "HackRF, Product Home Page," last modified February 29, 2024, https://greatscottgadgets.com/hackrf/one/

33 "USRP, Product Home Page," last modified February 29, 2024, https://www.ettus.com/

34 "BladeRF, Product Home Page," last modified February 29, 2024, https://www.nuand.com/

35 Maxim Integrated. (2015). "MAX2837 2.3GHz to 2.7GHz Wireless Broadband RF Transceiver," Product Datasheet.

36 Maxim Integrated. (2003). "MAX5864 Ultra-Low-Power, High-DynamicPerformance, 22Msps Analog Front End," Product Datasheet.

37 Silicon Laboratories Inc. (2011). "Si5351A/B/C I2C-Programmable Any-Frequency CMOS Clock Generator + VCXO," Product Datasheet.

38 Xilinx. (2008). "CoolRunner-II CPLD Family," Product Datasheet.

39 NXP Semiconductors. (2016). "LPC4350/30/20/10 32-bit ARM Cortex-M4/M0 Flashless MCU," Product Datasheet.

40 RFMD (2014). "*85–4200 MHz Wideband Synthesizer/VCO with Integrated 6 GHz RF Mixer*," RFMD.

5

Antennas for Handheld Wireless Devices

5.1 Introduction

This chapter focuses on antennas for handheld wireless devices. As the demand for wireless communication continues to grow, the importance of efficient and reliable antennas in handheld devices is increasingly emphasized. The role played by antennas in enabling the transmission and reception of signals in wireless devices cannot be understated, thus allowing users to remain connected and communicate seamlessly. In this chapter, we will explore the fundamentals of antennas, starting with a discussion on the radiation mechanism. A clear understanding of how electromagnetic waves are radiated by antennas is essential in designing and optimizing their performance for handheld devices. The sources of radiation and the characteristics that define the efficiency and effectiveness of antennas will be examined.

To ensure optimal functionality, careful consideration must be given to the parameters and specifications of antennas. The key parameters that impact antenna performance, such as gain, bandwidth, and impedance, will be explored. By understanding these parameters, informed decisions can be made when selecting and designing antennas for handheld wireless devices. The chapter will also cover different types of antennas commonly utilized in handheld devices and their corresponding performance characteristics. Various antenna designs, including monopole antennas and patch antennas, among others, will be discussed. By exploring the strengths and limitations of each antenna type, readers will gain insights into the selection of the most suitable antenna for specific applications.

Finally, practical design considerations for antennas in handheld wireless devices will be addressed. Antenna design poses challenges due to size and form factor constraints inherent in portable devices. Techniques and strategies to overcome these limitations while maintaining optimal performance will be discussed.

Introduction to Wireless System Design: From Circuits to Web-based Applications, First Edition.
Henry Lau, Ludy Liu, and Keith Chan.

5.2 Antenna Fundamentals

Antennas play a crucial role in wireless communication systems by determining their range and coverage. The design and characteristics of an antenna directly impact how far a signal can effectively be transmitted and received. By selecting appropriate antenna types and configurations, communication systems can be optimized to provide reliable connectivity over large areas or specific zones. Understanding the radiation mechanism is essential in developing suitable antennas for specific applications.

An antenna can be described as a structure that facilitates the transition between a guided wave and a free-space wave, as depicted in Figure 5.1 [1, 2]. This transition allows the conversion of radio frequency (RF) or microwave energy into electromagnetic waves that can propagate through space. The guided wave is typically transmitted through a transmission line, such as a 2-wire transmission line, coaxial cable, or waveguide. The transmission line serves the purpose of guiding the energy from one location to another with minimal losses.

Losses in the transmission line can include factors such as attenuation (the decrease in signal strength over distance), heat dissipation, radiation (energy loss in the form of electromagnetic waves), and mismatch loss caused by impedance mismatches. The transmission line aims to confine the transmitted energy within itself, ensuring efficient transfer to the antenna.

In Figure 5.1, the generator sends a uniform traveling wave along the transmission line. Upon reaching the open ends, the wave is radiated as a free-space wave. The wave guided within the transmission line is initially a plane wave but transforms into a spherically expanding wave at the open ends. Although the currents flowing from the generator through the transmission line cease at both ends, the fields generated by these currents continue to propagate. The region where the transition occurs from the guided wave to the free-space wave is referred to as the antenna.

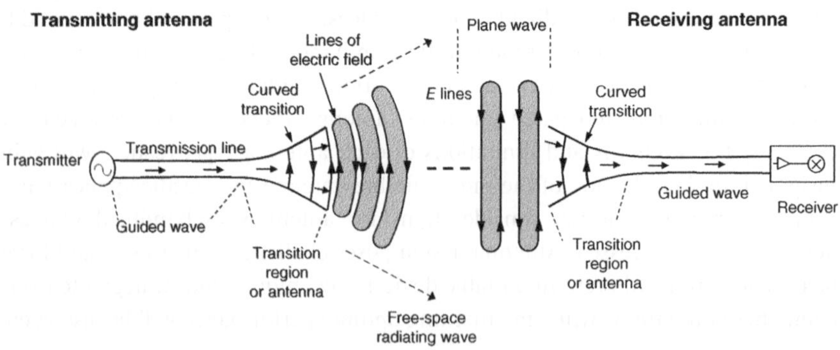

Figure 5.1 Antenna structure

Consequently, the antenna can be perceived as a device that converts currents from a circuit into electromagnetic photons that are emitted into free space. Similarly, as shown in Figure 5.1, the antenna can also function as a receiving device, converting photons back into currents directed towards the load.

Overall, antennas are essential components in wireless communication systems as they enable the efficient transmission and reception of signals, and the selection of appropriate antennas and understanding of their radiation mechanisms are crucial in achieving reliable and effective wireless connectivity.

5.2.1 Source of Radiation

For a more comprehensive understanding of how radiation is generated by the antenna, it is recommended to examine the movement of electrons within the conductor [3].

When electric charges undergo acceleration or deceleration, assuming the presence of a pulse of charge moving along a straight conductor in the x-direction, this moving charge instantaneously generates an electric current I (see Figure 5.2).

$$I = q_l \frac{dx}{dt} \tag{5.1}$$

where q_l is charge per unit length, C/m, and
C is in Coulomb.

Multiplying by the length l of the pulse along the conductor gives the current as

$$Il = q_l l \frac{dx}{dt} = qv \,(\text{A m}) \tag{5.2}$$

where v is velocity (m/s),
and
$q = q_l l$ = total charge of pulse (C).

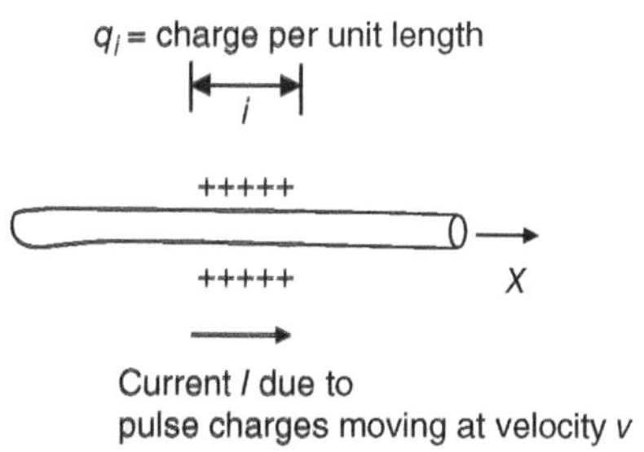

Figure 5.2 Movement of electrons within the conductor

For time-varying current, its time derivative gives

$$\frac{dI}{dt}l = q_l l\frac{d^2x}{dt^2} = q\dot{v}\ (\mathrm{A\,m/s}) \tag{5.3}$$

where $\dot{v}$ is the acceleration of charge (m/s^2)
Or alternatively,

$$\dot{I}l = q\dot{v}\ (\mathrm{A\,m/s}) \tag{5.4}$$

where $\dot{I}$ is the time-varying current (A/s),
$\dot{v}$ is the acceleration (m/s^2),
l is the length of current element (m), and
q is the electric charge (C).

This is the basic continuity relation between current and charge for electromagnetic radiation. From reference [3], it can be shown that the accelerated charge $q\dot{v}$ produces radiation, and the power P radiated is

$$P = \frac{\mu^2 q^2 \dot{v}^2}{6\pi Z}\ (\mathrm{W}) \tag{5.5}$$

where μ is the permeability of the medium (H/m).

It follows from Equation (5.4) and (5.5) that time-changing current produces radiation.

Therefore, it is clear that for radiation to occur from an antenna, there must be a conductor carrying a time-varying current that undergoes acceleration.

5.2.2 Characteristic of Radiation

To illustrate the characteristics of antenna radiation, Figure 5.3 shows a classical dipole antenna being fed by a transmission line. A dipole antenna is a widely used type of radio antenna for transmitting or receiving electromagnetic signals. It is a balanced antenna, meaning that it consists of two equal and opposite conductive

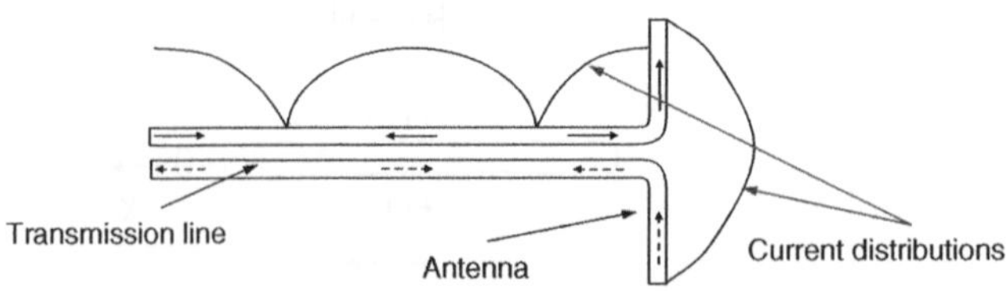

Figure 5.3 Source of antenna radiation

elements, typically in the form of straight wires or rods. These two conductive elements are oriented in parallel and are usually separated by a small gap.

One end of the transmission line is connected to the electric circuit, which serves as the output of the power amplifier for a transmitter. The electrical signal from the source travels along the transmission line towards the ends of the dipole and then bounces back. In the case of a sinusoidal signal source, the current takes the form of a standing wave, depending on factors such as the length of the transmission line and the dipole, as well as the impedance of the signal source and the antenna. As previously explained, it is the distribution of current along the dipole antenna that produces the radiation.

5.3 Parameters and Specifications

5.3.1 Radiation Resistance

One of the key parameters of an antenna is its radiated power which can be related to the radiation resistance [1, 2].

The radiation resistance of an antenna can be defined as the theoretical resistance that would dissipate an equivalent amount of power as is radiated by the antenna, as illustrated in Figure 5.4. It is an imaginary resistance that, when connected in series with the antenna, would consume the same power as the actual radiation. To achieve maximum radiation power, it is important to design the antenna in a way that maximizes its radiation resistance.

This concept can be illustrated by considering the example of a dipole antenna discussed in the previous Chapter. The radiation resistance of the dipole can be determined by applying the principles of electromagnetic theory, particularly

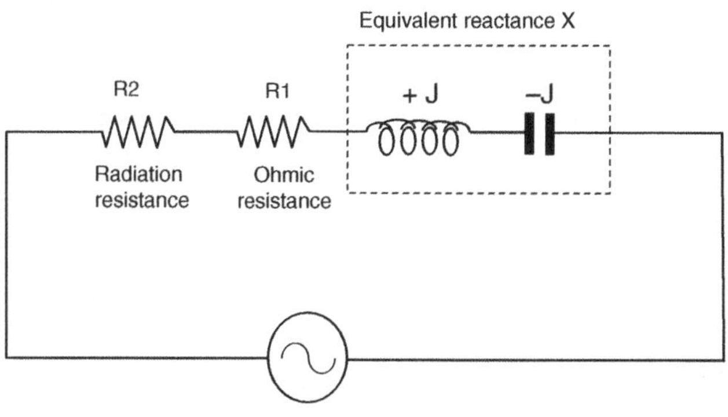

Figure 5.4 Equivalent model of radiation resistance

the Poynting theorem [4], which is related to the power radiated. In the case of a short dipole with a length of 2 h, the radiation impedance can be expressed as follows:

$$R_{\text{rad}} = 80\pi^2 \left(\frac{h}{\lambda}\right)^2 \tag{5.6}$$

As demonstrated by Equation (5.6), there is a direct correlation between the length of the wire and the radiation power, with longer wires resulting in higher radiation resistance and consequently increased power. When the length of the dipole antenna is half the wavelength, the corresponding radiation resistance is 73 Ω.

5.3.2 Efficiency

Efficiency [1, 2] is a crucial factor when discussing antennas, especially small ones, as it quantifies the ratio of power radiated from the antenna to the power dissipated in the antenna structure, resulting in heat.

The efficiency is calculated using the equation:

$$\eta = Rr/(Rr + Rd) \tag{5.7}$$

In the equation, Rr denotes the radiation resistance (desired), while Rd represents the dissipation resistance (undesired). For small antennas, which typically have low radiation resistance, any circuit loss (resistance) results in a significant reduction in efficiency. This effect is particularly pronounced in small magnetic loop antennas.

5.3.3 Directivity

The radiation pattern is described by the antenna's directivity [1, 2, 4], which determines how effectively the antenna can "see" and radiate in different directions. In a fixed point-to-point radio link, the directivity can be utilized to concentrate the radiation beam in the desired direction. However, in systems where the transmitter and receiver positions are not fixed, a balanced radiation distribution in all directions is sought. In other words, an omnidirectional antenna is preferred to ensure equal radiation coverage.

5.3.4 Polarization

Polarization [1, 2, 4] is a descriptor of the direction of the electric field in an antenna. In free space, electromagnetic waves exhibit electric and magnetic fields that are perpendicular to the direction of propagation. Typically, when discussing polarization, emphasis is placed on the electric field vector, while the magnetic field is disregarded due to its perpendicularity to the electric field. Optimal performance requires both the receiving and transmitting antennas to possess matching polarization.

In practical handheld applications, antennas often produce fields with polarization in multiple directions, and reflections can further alter the polarization of the electric field. Consequently, polarization is less critical for indoor equipment, which encounters numerous reflections, compared to outdoor equipment operating with a line of sight. Certain antennas generate electrical fields with predetermined directions, necessitating knowledge of the polarization utilized during radiation pattern measurements. Additionally, specifying the frequency at which the measurements were conducted is crucial. Generally, radiation patterns exhibit minimal changes across frequencies, making it customary to measure them at the midpoint of the antenna's designated frequency band. Although the relative level may slightly fluctuate within the desired frequency range for narrowband antennas, the fundamental shape of the radiation pattern remains largely unaltered.

5.3.5 Antenna Gain

The antenna gain [1, 2, 4], or simply gain, is another key parameter that describes the performance of an antenna. It is related to both the electrical efficiency and the directional property of the antenna. The effectiveness of the antenna in turning its input power into a radio signal pointed in a particular direction is determined by its gain as a transmitting antenna. Similarly, the effectiveness of the antenna in converting incoming radio signals into electrical signals is described by its gain as a receiving antenna.

Typically, the gain is calculated in the direction of maximum radiation.

The gain is given by the formula:

$$G = D \times \eta \times m \tag{5.8}$$

where D represents the directivity,
η represents the efficiency,
and m represents the mismatch loss.

The gain is usually expressed in units of dBi or dBd. In the case of dBi, the directivity (D) is measured relative to an isotropic radiator, which has equal radiation in all directions. dBd is used when the directivity refers to a dipole antenna.

A gain above that of an isotropic radiator, expressed as dBd = 2.15 dBi, is exhibited by a dipole antenna itself.

5.3.6 Effective Isotropic Radiated Power (EIRP)

The Effective Isotropic Radiated Power (EIRP) [1, 2, 4] refers to the effective power radiated from an antenna, taking into consideration the antenna's gain. Maximum output power measurements in regulatory specifications often make use of EIRP as a reference.

$$EIRP = G \times P \tag{5.9}$$

Here, G represents the antenna gain, and P represents the output power from the transmitter.

5.4 Types of Antenna and Performance

5.4.1 Basic Antennas

5.4.1.1 Dipole

As mentioned previously, a dipole antenna [1, 2] is a specific type of radio antenna composed of two conductive elements, typically wire rods or metal rods, arranged in a straight line with a small gap between them, as depicted in Figure 5.5. The term "dipole" signifies that the antenna possesses two poles, situated at each end of the conductive elements. Dipoles are essential components in both practical applications and the mathematical analysis of antennas.

When the length of a linear conductor is significantly shorter than the wavelength ($L << \lambda$), it is commonly referred to as a short dipole. The power radiated by the antenna is generally proportional to its length (L), implying that a longer antenna will exhibit better radiation performance. For optimal radiation efficiency, the length of the dipole is half the wavelength. It is crucial to position the antenna away from the ground plane and any metallic or conductive objects. The dipole antenna has an impedance of approximately 73 Ω, making it easily matchable to 50 Ω. Additionally, the dipole antenna has a gain of 2.14 dBi. The polarization of the electromagnetic field corresponds to the orientation of the antenna element, which can be either horizontal or vertical.

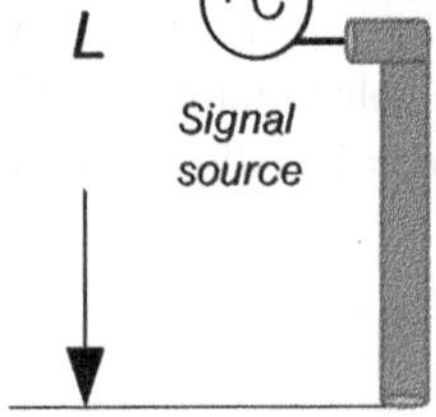

Figure 5.5 Dipole antenna

Despite its advantages, a longer antenna would require more space, which is often limited in wireless handheld devices. Since a dipole antenna is differential in nature, it requires either a differential signal source or a balun to convert a single-ended source into a differential one.

5.4.1.2 Monopole

The monopole antenna [1, 2], commonly referred to as the whip antenna, is essentially a 1/4 wavelength wire positioned above a ground plane for optimum radiation

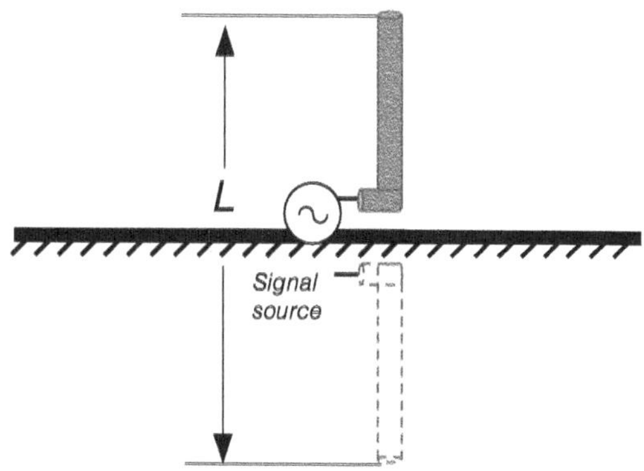

Figure 5.6 Monopole antenna

efficiency, as shown in Figure 5.6. The antenna operates by utilizing the presence of the ground plane, and it is recommended to keep the "far" open end away from it. The monopole antenna is fed single-ended and exhibits an impedance of 37 Ω at the resonance frequency, making it easily matched to a 50 Ω impedance.

Designing monopole antennas is a straightforward process, allowing for fine adjustments to the antenna length to modify its resonance frequency. Like the dipole antenna, the monopole antenna can have either horizontal or vertical polarization, depending on its orientation. Additionally, the monopole antenna's radiation pattern closely resembles that of the dipole antenna, and theoretically, its gain is 3 dB higher than that of the dipole antenna due to the power being radiated solely in the upper half plane, thanks to the ground plane. The monopole antenna is considered the optimal solution when the physical size is acceptable and a ground plane is available. However, utilizing a smaller ground plane will impact the antenna's performance and cause the radiation pattern to tilt.

5.4.1.3 Loop Antenna

The loop antenna [1–2, 5–6] can be considered as a large coil and is typically tuned to parallel resonance at the desired frequency by adding a parallel capacitor, as shown in Figure 5.7. By adding a parallel capacitor, the loop antenna's input impedance is significantly increased, measured in kilohms. A small loop exhibits a very narrow bandwidth and has a high Q factor. This characteristic provides an advantage for the selectivity of the device but also makes the tuning process extremely critical. However, once tuned, the loop antenna can still be easily affected by hand-induced detuning. Furthermore, the loop antenna radiates a magnetic field, which enhances its performance when positioned in close proximity to the body. This feature makes it a suitable solution for applications such as pagers,

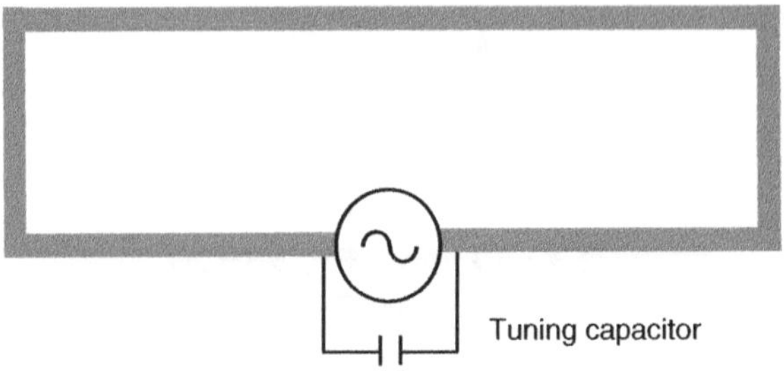

Figure 5.7 Loop antenna

particularly those involving remote or on-body usage. The polarization of the loop antenna aligns with its surface, resulting in an omnidirectional radiation pattern.

The most significant disadvantage of loop antennas, when compared to the two previously mentioned antenna types, is their poor efficiency particularly at low frequency. This is due to the fact that the radiation resistor has a much lower value than the loss resistor. The gain of the loop antenna is dependent on its size, although a realistic expected gain in practical scenarios is approximately −10 dBi. Similar to the Printed Circuit Board (PCB) antenna to be discussed later in this Chapter, the integration of the loop antenna into the PCB can yield a low-cost solution. The antenna tuning process often involves the use of a variable capacitor, which contributes to the overall cost. However, it may be practical to utilize a nonvariable capacitor, although careful adjustment during the engineering stages is necessary to ensure proper tuning with a standard value capacitor.

5.4.1.4 Patch Antenna

A microstrip antenna, also referred to as a patch antenna [2, 7], is a compact and lightweight antenna widely used in wireless communication devices. It is specifically designed for high-frequency operation, typically spanning from several hundred Megahertz to several Gigahertz. Patch antennas have diverse applications across several fields, including wireless communication devices, satellite communication systems, radar systems, and RFID systems. The antenna structure comprises a metallic patch, usually composed of conductive materials like copper or silver, that is printed on a dielectric substrate. This configuration is illustrated in Figure 5.8. The dielectric substrate is commonly made of ceramic material, possessing a dielectric constant ranging from 10 to 80. This characteristic enables the antenna to maintain its effectiveness while occupying a smaller physical footprint. The patch itself is typically rectangular or circular in shape and is mounted on a ground plane, which acts as a conductive surface for the antenna.

When an electrical signal is applied to the patch, it generates electromagnetic waves that are emitted into space. The size, shape, and composition of the patch,

Figure 5.8 Patch antennas in different sizes

in conjunction with the substrate properties, determine the antenna's operating frequency and performance characteristics. By carefully optimizing these parameters, the antenna can be precisely tuned to operate at a specific frequency or within a desired frequency range. Patch antennas offer numerous advantages. They possess a compact and lightweight design, making them cost-effective to manufacture due to their simple structure. Furthermore, their form factor allows for convenient integration into printed circuit boards or flat surfaces, rendering them suitable for handheld devices and applications with space constraints. Additionally, patch antennas exhibit relatively wide bandwidth and high radiation efficiency, making them particularly well-suited for wireless communication systems that necessitate high data rates and long-range communication.

5.4.2 Miniature Antennas

5.4.2.1 Chip Antenna

Chip antennas [8, 9] can be considered a good solution when there is limited board space available for the antenna. Even for frequencies below 1 GHz, small-sized solutions can be achieved due to the high dielectric constant of the antenna material, typically ceramic. However, compared to PCB antennas, using chip antennas requires adding a component to the Bill of Materials and incurring mounting costs. The cost of a typical chip antenna ranges between 0.10 and 0.50 USD.

Although manufacturers claim that chip antennas are matched to 50 Ω for a specific frequency band, it is often necessary to use additional matching components to achieve optimal performance. The performance numbers and recommended matching provided in data sheets are typically based on measurements conducted with a test board. The dimensions of this test board are usually documented in the data sheet. It is important to note that the performance and required matching will vary if the chip antenna is implemented on a PCB with a different size, shape, and material for the ground plane.

A chip antenna, also known as a surface-mount antenna or planar antenna, is a compact antenna designed to be directly mounted onto a PCB. It is called a "chip" antenna because of its small size and resemblance to an electronic chip. Figure 5.9 illustrates a chip antenna used in a Bluetooth (BLE) beacon device, which occupies very little board space. Chip antennas are used for wireless communication in various devices such as smartphones, tablets, Wi-Fi routers, Bluetooth devices, and other wireless-enabled devices. They are widely used due to their small size, low profile, and ease of integration into PCB designs. The design of a chip antenna involves etching or printing the antenna structure onto a dielectric substrate material, such as ceramic or organic material. The antenna structure is often a meandering trace or a loop structure, and it can be designed for specific frequency bands or multiple bands depending on the application.

The advantages of chip antennas include their compact size, which allows for space-saving integration into small electronic devices, ease of manufacturing and assembly, and comparatively low cost. However, chip antennas also have some limitations, such as lower efficiency and narrower bandwidth compared to larger, external antennas. The performance of a chip antenna can be affected by factors such as the proximity of other components on the PCB, the ground plane, and surrounding structures. Overall, chip antennas are a popular choice for wireless communication applications where space is limited, and moderate performance is acceptable. Their small size and compatibility with PCB manufacturing processes make them a convenient option for integrating wireless capabilities into various electronic devices.

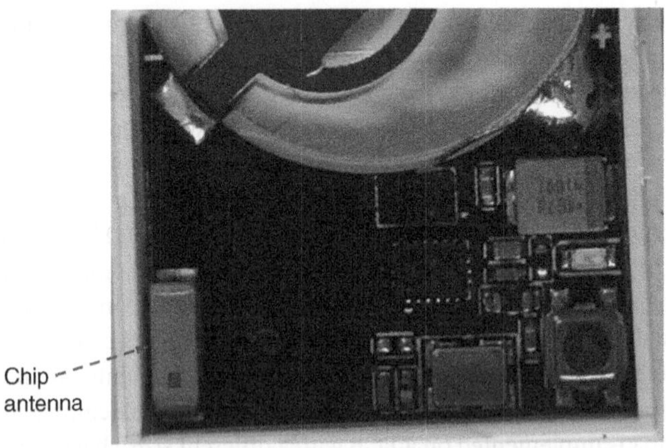

Figure 5.9 Ceramic chip antenna on a BLE PCB

5.4.2.2 Inverted-L Antenna

Bending and folding techniques are widely employed for antenna miniaturization, offering a practical solution to reduce the size of dipole or monopole antennas. Although dipole and monopole antennas exhibit high antenna efficiency, they typically have long lengths. They are thus commonly used as external antennas since it would be very difficult to house them if used internally. To achieve a smaller footprint without compromising performance, a popular approach involves bending the top portion of a monopole antenna to create an inverted shape, known as an inverted-L antenna, as depicted in Figure 5.10.

Inverted-L antennas are often used in limited space environments or where height restrictions exist. By adopting this configuration, the inverted-L antenna occupies significantly less space compared to a traditional monopole, as long as there is sufficient horizontal axis room available. They offer both vertical and horizontal polarization and can provide efficient radiation patterns. However, they are generally less efficient than taller vertical antennas and may have reduced gain and bandwidth.

It is necessary to use a matching network to convert the radiation impedance of a quarter-wavelength monopole, typically around 36.5 Ω, to a system impedance of 50 Ω for numerous applications. When employing an inverted-L antenna, impedance transformation becomes crucial due to its lower radiation resistance resulting from the ground plane effect [2, 10]. Additionally, the capacitance of the inverted-L antenna is higher compared to a monopole, as the bent segment is in closer proximity to the ground plane, leading to increased capacitive coupling. However, the inclusion of a matching network unavoidably introduces RF signal attenuation due to losses associated with the matching elements.

For frequencies below 3 GHz, surface-mounted capacitors and inductors are frequently employed in matching networks. Surface-mounted capacitors primarily experience dielectric losses at high frequencies, while the losses in inductors depend on the material and can be either dielectric or ferrite losses [11]. On the other hand, for frequencies above 3 GHz, distributed elements are more suitable for matching, resulting in lower losses. However, they require more physical space compared to surface-mounted components. The ideal scenario is to design the antenna in a manner where its sectional elements serve both as matching and

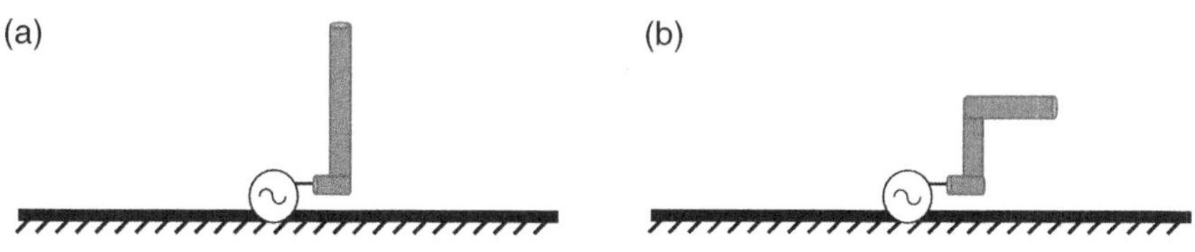

Figure 5.10 (a) A monopole antenna (b) An Inverted-L antenna showing the bent section

radiating elements simultaneously. This eliminates the need for external matching elements while maintaining high efficiency.

5.4.2.3 Inverted-F Antenna

The inverted-F antenna [12, 13] shown in Figure 5.11 serves as an excellent example of designing an antenna where its sectional elements perform dual functions as both matching and radiating elements. It bears similarities to an inverted-L antenna but comprises three segments: an upper arm, a feeding conductor, and a shunting arm. The upper arm, bent along the x-axis, acts as the primary radiating segment, while the shunting arm diverts a portion of the signal to ground, with the signal being injected through the feeding conductor.

Conceptually, the inverted-F antenna can be viewed as an open-ended transmission line as depicted in Figure 5.12. The impedance at the open end typically exhibits high values and can be represented as an equivalent impedance of Rant || jXant, characterized by maximum voltage and zero current. Conversely, the impedance at the short end is zero, with maximum current and zero voltage. As the feeding point progresses from the short end towards the open end, the impedance undergoes a transition from very low to very high values. At a specific location along the transmission line (antenna), the impedance closely matches the system impedance of 50 Ω. The transmission line formed by the antenna's conductor serves the dual purpose of both matching and radiating elements.

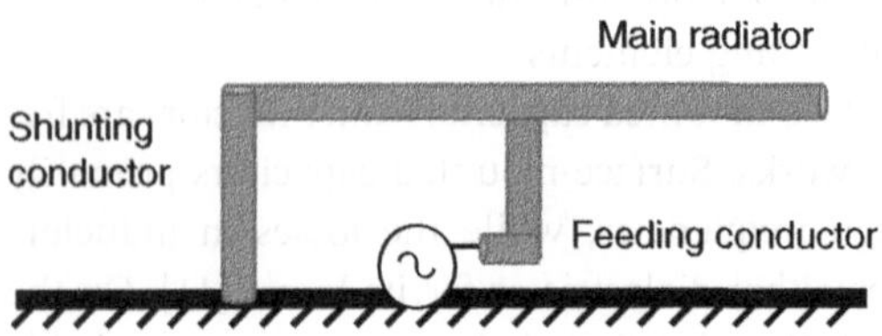

Figure 5.11 Inverted-F antenna

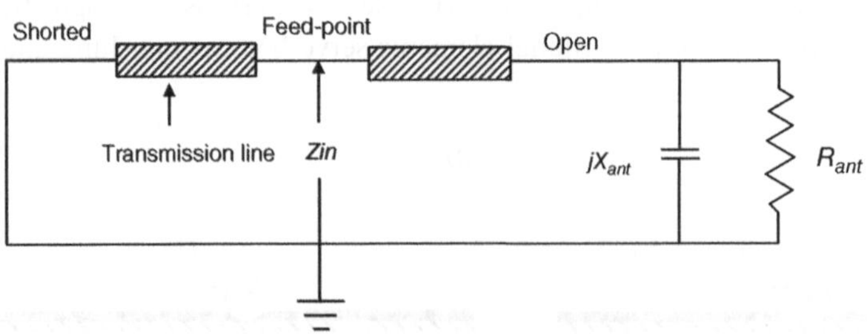

Figure 5.12 Transmission line model of an Inverted-F antenna

5.4.2.4 Meandered Line Antennas

Meandered line antennas are primarily developed for miniaturization and extensively used in small handheld devices due to their compact sizes [14–16]. A typical meander line dipole consists of periodic segments of vertical and horizontal branches. When compared to a straight dipole of the same height, the meander line dipole has a longer physical length, resulting in a lower resonant frequency. The reduction in resonant frequency achieved through meandering is due to the increased physical length of the dipole. This allows for antenna size reduction while maintaining the same operating frequency, making meander line antennas highly suitable for small handheld devices. Passive UHF RFID tags extensively employ this approach for their antennas, as shown in Figure 5.13.

The radiation efficiency drops with an increase in the number of turns due to the cancellation between segments with opposite signal phases [17], despite the shrinkage in the physical size of the antenna. Therefore, optimizing the geometry of the meander line antenna is necessary to achieve optimum performance in terms of size, efficiency, bandwidth, and input impedance.

5.4.3 Antenna Array

An antenna array [2, 4, 18], also known as a phased array, is a collection of multiple individual antennas that work together to enhance the performance of a wireless communication system. Instead of using a single antenna, an antenna array employs multiple antennas, which are spatially arranged in a specific configuration. The antennas in an array can be physically separated by a certain distance or placed in close proximity to each other. They are connected to a signal processing system that controls the phase and amplitude of the signals transmitted or received

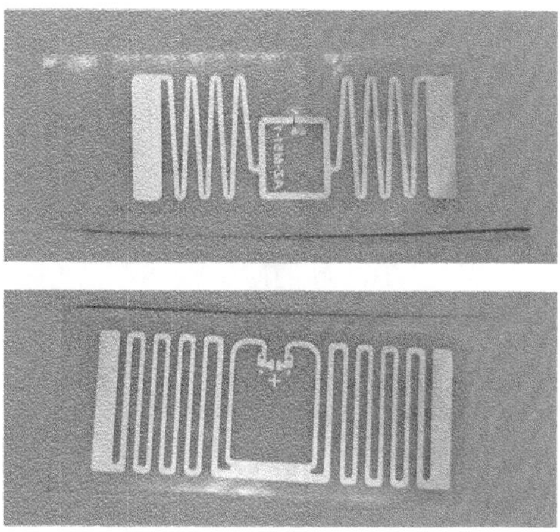

Figure 5.13 Meandered line antennas

by each individual antenna. By manipulating the phase and amplitude of the signals, the antenna array can shape and steer the radiation pattern in a desired direction.

One of the key advantages of antenna arrays is their ability to perform beamforming. By adjusting the phase and amplitude of the signals at each antenna element, the array can create constructive interference in a specific direction, thereby focusing the transmitted or received energy in that direction. This allows for improved signal strength, increased range, and reduced interference. A typical antenna array example is the collinear antenna array which is an array of dipole mounted in such a manner that the corresponding elements of each antenna are parallel and collinear; that is, they are located along a common axis as shown in Figure 5.14.

Collinear arrays are high gain omnidirectional antennas. Both dipoles and quarter-wavelength monopoles have an omnidirectional radiation pattern in free space when oriented vertically; they radiate equal radio power in all azimuthal directions perpendicular to the antenna, with the signal strength dropping to zero on the antenna axis. The purpose of stacking multiple antennas in a vertical collinear array is to increase the power radiated in horizontal directions and reduce the power radiated into the sky or down towards the earth, where it is wasted. They radiate vertically polarized radio waves. Theoretically, when stacking idealized lossless antennas in such a fashion, doubling their number will produce double the gain, with an increase of 3 dB. In practice, the gain realized will be below this

Figure 5.14 Comlinear antenna array. *Source:* Sanjo/Wikimedia Commons/CC BY-SA 4.0

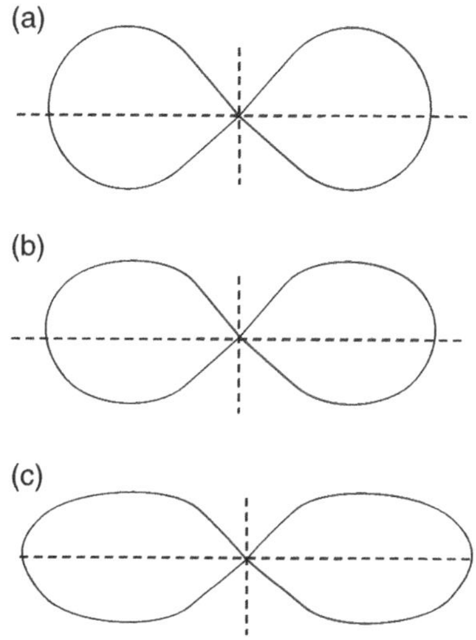

Figure 5.15 Radiation pattern for different number of array elements (a) single element (b) two elements, and (c) three elements

due to imperfect radiation spread and losses. Figure 5.15 shows the increase in gain with increasing number of array elements.

An example of a typical antenna array is the collinear antenna array. It consists of a series of dipole antennas arranged in a parallel and collinear fashion, meaning they are positioned along a common axis, as depicted in Figure 5.14. Collinear arrays are known for their high gain and omnidirectional characteristics. When dipole or quarter-wavelength monopole antennas are oriented vertically, they exhibit an omnidirectional radiation pattern in free space. This means they emit radio power equally in all directions perpendicular to the antenna, except along the antenna axis where the signal strength drops to zero. By stacking multiple antennas in a vertical collinear array, the aim is to maximize the power radiated horizontally while minimizing wasteful radiation towards the sky or the ground. Collinear arrays emit vertically polarized radio waves. Theoretically, when ideal lossless antennas are stacked in this manner, doubling the number of antennas results in double the gain, with a 3 dB increase. However, in practical scenarios, the realized gain is lower than this theoretical value due to imperfect radiation distribution and losses. Figure 5.15 illustrates how the gain increases with the growing number of array elements.

Antenna arrays also provide spatial diversity by exploiting the multipath propagation characteristics of wireless signals. By using multiple antennas, the array can receive signals from different paths, mitigating the effects of fading, signal attenuation, and interference. This results in improved signal quality and increased reliability. Furthermore, antenna arrays can dynamically adapt their radiation pattern based on the changing conditions of the wireless channel. By continuously monitoring the channel and adjusting the phase and amplitude of the signals, the array can track and maintain optimal signal reception or transmission even in a dynamic environment. This adaptive beamforming capability further enhances the overall performance of the communication system.

Antenna arrays are utilized in a wide range of fields, such as radar systems, satellite communications, wireless communication networks, and wireless sensor networks. These arrays bring notable benefits in terms of performance, flexibility, and adaptability when compared to single-antenna systems.

5.5 Practical Design Considerations for Handheld Wireless Devices

Handheld wireless devices have become integral parts of our lives, offering connectivity, information access, productivity tools, entertainment, and much more, including smartphones, tablets, handheld gaming consoles, e-book readers, wearable devices, portable GPS navigation systems, wireless earphones, and more; their importance lies in their ability to enhance communication, productivity, convenience, and the overall quality of life in an increasingly connected world.

Designing antennas for handheld wireless devices involves several practical considerations and techniques. One important factor is the size and form factor of the antenna, as space is often limited in portable devices. Miniaturization techniques, such as using small printed or chip antennas, or integrating antennas into the device's structure, help achieve compact designs.

Frequency band operation is crucial in antenna design as it determines the size and shape of the antennas. Antennas need to be specifically designed to operate in the desired frequency range. The characteristics of the frequency band dictate the optimal dimensions and configuration of the antennas to ensure efficient transmission and reception of signals. Cellular networks, such as 3G, 4G, and 5G, operate on specific frequency bands that are allocated by regulatory authorities. These frequency bands can vary across different countries and regions. Regulatory bodies assign different frequency ranges to cellular networks to avoid interference and ensure efficient utilization of the radio spectrum. For cellular phones to connect to the network infrastructure, they must support multiple frequencies.

By supporting multiple frequencies, cellular phones can communicate with the available network bands. This enables users to access cellular services and enjoy functionalities like voice calls, messaging, Internet browsing, and data-intensive applications. Support for multiple frequency bands is often required in portable electronics. Antenna designs that can operate across multiple bands, such as multiband or reconfigurable antennas, provide broad frequency coverage. Efficient transmission and reception of signals require optimization of radiation patterns. Antennas are designed to emit and receive signals in specific directions to minimize signal loss and interference. By optimizing radiation patterns, cellular phones can effectively transmit and receive signals in the intended directions, improving signal quality and overall performance.

Impedance matching is vital for maximizing power transfer and minimizing signal reflections. Techniques like matching networks or impedance matching components [19–21] are used to achieve good impedance matching between the antenna and the device's circuitry. The design of the PCB layout, particularly the ground plane beneath the antenna, is critical to minimize interference and maintain proper antenna characteristics. Careful placement and integration of the antenna within the device are necessary to avoid interference or obstruction by other components.

Advanced materials play a key role in miniaturized antenna performance. Specialized materials with unique electromagnetic properties, such as high-permittivity substrates or metamaterials [22–24], are used to enhance efficiency and control radiation patterns. Integration with other device features is important, considering the coexistence of antennas with components like displays, buttons, or camera modules. Design considerations should minimize interference or detuning caused by nearby components. Compliance with regulatory requirements and standards, including specific absorption rate (SAR) [25] limits and electromagnetic compatibility regulations, is essential.

To optimize antenna performance, simulation tools, prototyping, and testing are often employed, with iterative design improvements. Collaboration between antenna designers, RF engineers, mechanical, and product designers is crucial for successful antenna integration into handheld wireless devices.

References

1 Kraus, J.D. (1988). *Antennas*. New York: McGraw-Hill.

2 Balanis, C.A. (1997). *Antenna Theory - Analysis and Design*. John Wiley & Sons Inc.

3 Jordan, E.C. and Balmain, K.G. (1968). *Electromagnetic Waves & Radiating Systems*, 2e. Prentice Hall.

4 Stutzman, W.L. and Thiele, G.A. (2012). *Antenna Theory and Design*. John Wiley & Sons.

5 Smith, G.N. (1972). Radiation efficiency of electrically small multiturn loop antennas. *IEEE Transactions on Antennas and Propagation*, AP-20 (5): 656–657.

6 Lau, H. (1995). "Performance concerns guide the design of pager antennas," *Microwaves & RF*, pp. 95–106.

7 Pozar, D.M. and Schaubert, D.H. (1995). *Microstrip Antennas, the Analysis and Design of Microstrip Antennas and Arrays*. John Wiley & Sons Inc.

8 Ray PCB & Assembly, "What is a Chip Antenna and How Does it Work?," last modified January 25, 2024, https://www.raypcb.com/chip-antenna/

9 Johanson Technology, "RF Ceramic Chip Antennas," last modified January 25, 2024, https://www.digikey.hk/zh/ptm/j/johanson-technology/rf-ceramic-chip-antennas/tutorial

10 Volakis, J., Chen, C.-C., and Fujimoto, K. (2009). *Small Antennas: Miniaturization Techniques & Applications*, 1e. McGraw-Hill.

11 Rashed, J. and Tai, C.T. (1991). "A new class of resonant antennas," *IEEE Transactions on Antennas and Propagation* 39: 1428–1430.

12 Zhu, Q., Fu, K., and Li, Y. (2005). "The analysis of planar inverted-F antennas with equivalent models," *Journal of University of Science and Technology of China* 35 (2): 144–147.

13 Gallo, M., et al. (2011). "Design of an inverted F antenna by using a transmission line model," *Proceedings of the 5th European Conference on Antennas and Propagation*, pp. 635–637, April 2011.

14 Hui, L., et al. (2015). "A small printed antenna for bluetooth wireless communication," *Progress in Electromagnetics Research Symposium Proceedings*, pp. 2190–2194.

15 Fenwick, R.C. (1965). "A new class of electrically small antennas," *IEEE Transactions on Antennas and Propagation* 13: 379–383.

16 Rashed, J. and Tai, C. T. (1982). "A new class of wire antennas," *IEEE Antennas and Propagation Society International Symposium*, vol. 20, pp. 564–567.

17 Tsutomu, E. et al. (2000). "Resonant frequency and radiation efficiency of meander line antennas," *Electronics and Communications in Japan* 83: 52–58.

18 Mailloux, R.J. (2015). *Phased Array Antenna Handbook*. Artech House.

19 Jackson, H.W. (1959). *Introduction to Electronic Circuits*. Prentice-Hall.

20 "Maximum Power Transfer Matching," last modified January 25, 2024, https://en.wikipedia.org/wiki/Impedance_matching#Maximum_power_transfer_matching

21 Lau, H. (2019). *Practical Antenna Design for Wireless Products*. Artech House.

22 Volakis, J., Chen, C.-C., and Fujimoto, K. (2009). *Small Antennas: Miniaturization Techniques & Applications*, 1e. McGraw-Hill.

23 Acharya, R. et al. (2017). "A review on antenna application of metamaterials," *Journal of Electronics and Communication Engineering* 12 (5): 6–8.

24 Singh, G., Rajni, and Marwaha, A. (2015). "A review of metamaterials and its applications," *International Journal of Engineering Trends and Technology* 19 (6): 306–310.

25 International Commission on Non-Ionizing Radiation Protection, "Guidelines for Limiting Exposure to Electromagnetic Fields," last modified January 25, 2024, https://www.icnirp.org/en/publications/guidelines/index.html

6

PCB Design for Wireless Devices

6.1 Introduction

In the field of wireless technology, the design and development of printed circuit boards (PCBs) play a crucial role in ensuring optimal performance and functionality of wireless devices. PCBs designed for wireless devices require specialized knowledge and techniques to address the unique challenges posed by high-frequency signals, signal integrity, power distribution, and interference mitigation.

This chapter provides a comprehensive overview of the key considerations and best practices for designing PCBs specifically tailored for wireless applications. It covers various aspects of RF PCB design, including layer stack-up assignment, component placement, grounding methods and techniques, power plane design, and bypassing and decoupling strategies. This chapter starts by exploring the fundamental aspects of RF PCB design. It discusses the assignment of layer stack-up, considering factors such as the number and arrangement of PCB layers, in order to optimize signal integrity and provide better signal routing. Additionally, this chapter examines the strategic placement of components to minimize signal path lengths, reduce noise coupling, and enhance the Electromagnetic Interference/Electromagnetic Compatibility (EMI/EMC) performance of the RF circuit.

Grounding methods and techniques are then discussed, as they are crucial for maintaining signal integrity and mitigating interference. Grounding is essential in RF PCB design for signal integrity, EMC, and optimal component performance. It reduces noise, controls impedance, and shields RF circuits from external interference. Proper grounding minimizes ground loops, facilitates clear return current paths, and aids in thermal dissipation. Bypassing and decoupling techniques are addressed as essential elements of RF PCB design. The significance of bypass capacitors and decoupling networks in suppressing power supply noise and ensuring stable operation of RF circuits is discussed. Furthermore, this chapter extends the discussion to cover other important circuits in PCB design for wireless devices.

Introduction to Wireless System Design: From Circuits to Web-based Applications, First Edition.
Henry Lau, Ludy Liu, and Keith Chan.

It highlights the key differences in design considerations between PCBs for IF circuits and RF circuits. Strategies to help minimize noise interference and reduce crosstalk in baseband circuits are discussed, along with the unique challenges associated with designing audio circuits for wireless devices. Lastly, this chapter presents measures for PCB design to ensure reliable and efficient power supply operation.

By providing insights into these topics, this chapter aims to equip engineers and designers with the knowledge necessary to create high-performance PCB layouts tailored for wireless devices. The principles and techniques discussed cater to the demanding requirements of modern wireless applications.

6.2 RF PCB Design Fundamentals

When designing PCBs for wireless applications, it is important to consider certain basic principles and guidelines that can impact the performance and cost of the PCBs. These principles are the fundamentals of RF PCB design, which will be presented in the following sections.

6.2.1 Layer Stack-up Assignment

One of the main aspects of RF PCB design is the layer stack-up configuration. A layer stack-up determines how the different layers of the PCB are arranged and connected [1]. It affects the performance, reliability, and cost of the RF circuit. A common guideline for RF PCB stack-up design is to use at least a four-layer configuration with internal power and ground planes. This provides effective decoupling and power integrity for high-speed digital interfaces. Placing the power and ground planes in adjacent internal layers also reduces the loop area and the inductance of the power distribution network.

Another important aspect of RF PCB design is the signal layer placement. The signal layer is where the RF signals are routed and processed. It should be placed on the top or bottom layer, adjacent to a ground plane. This allows for 50 Ω impedance routing with narrow trace widths and complete isolation from the digital sections. The ground plane should have sufficient via stitching to provide a low-impedance return path for the RF signals.

Although a four-layer configuration is preferred, it may be necessary to reduce the number of layers due to cost considerations. A single-layer board design with surface-mounted (SMT) and lead-type components is shown in Figure 6.1. It is characterized by the lowest cost as it has only one copper layer. The lead-type component is placed on the top side, which serves as the component layer, while the copper layer is located on the bottom side. Furthermore, if the wave soldering

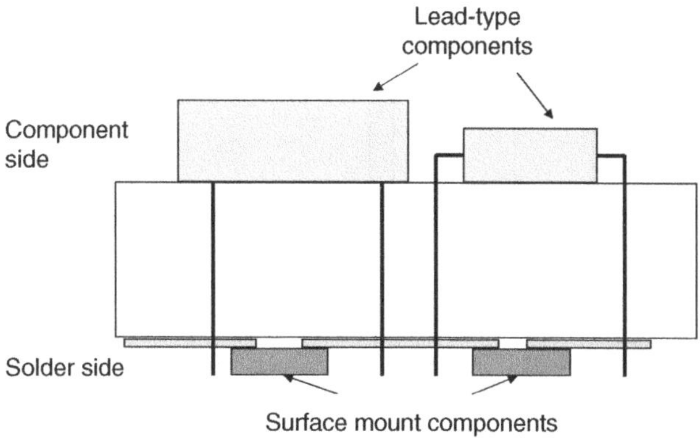

Figure 6.1 Single-layer PCB design

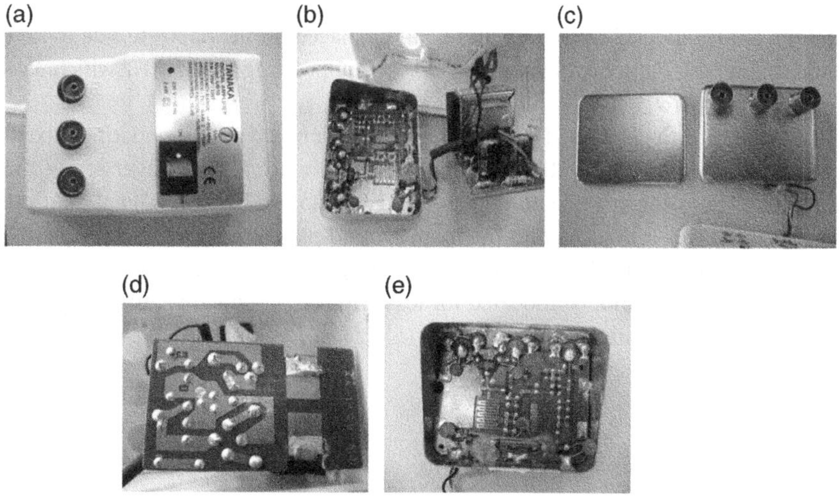

Figure 6.2 An analog TV signal booster (a) complete product, (b) RF amplifier and power supply boards, (c) RF amplifier in a shield box, (d) solder side of the power supply board, (e) SMT and lead-type components

method is used, both the SMT components and the copper layer can be soldered simultaneously. Figure 6.2 showcases a product, an analog TV signal booster, which utilizes single-layer boards. The product consists of two single-layer PCBs: the power supply board and the RF receiver board. The RF receiver board contains both SMT and lead-type components and utilizes the cost advantage offered by the wave soldering method.

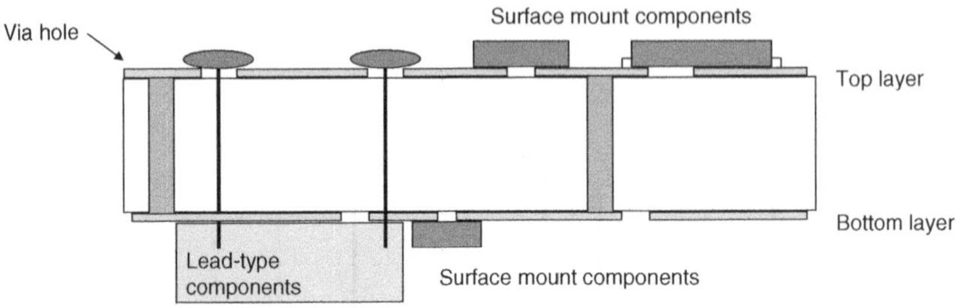

Figure 6.3 Two-layer board design

For a more complicate design with more components and signal tracings, a two-layer board may be required as shown in Figure 6.3. The top layer is usually used for component mounting and major signal routing while the bottom layer is primarily filled with ground plane and power traces. Similar to the single-layer design, SMT and lead-type component can be placed on the opposite side to save production cost. Figure 6.4 shows an RF receiver operating at 400 MHz which uses a two-layer board.

In Figure 6.5, the preferred design for a four-layer board accommodating mixed-signal analog, digital, and RF applications is illustrated. The top layer serves the purpose of housing major components and facilitating primary signal routing. The second layer is dedicated to the main ground plane and reference. Typically, the third layer is utilized for minor signal routing and the power plane. Lastly, the bottom layer is commonly allocated for less critical components, auxiliary signal routing, and the ground plane.

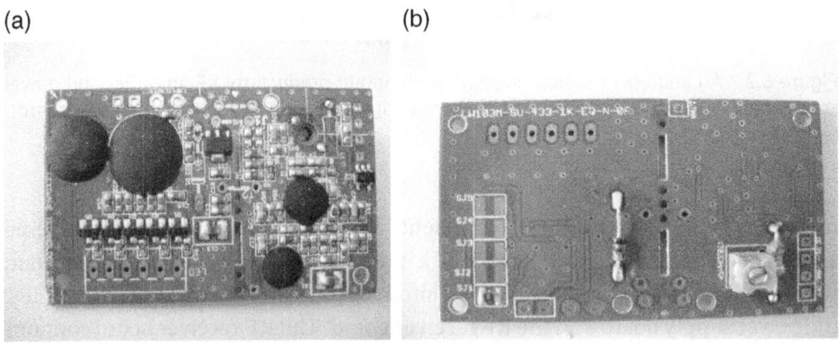

Figure 6.4 A 400 MHz receiver using a two-layer board (a) top side with SMT components and (b) bottom side with lead-type components

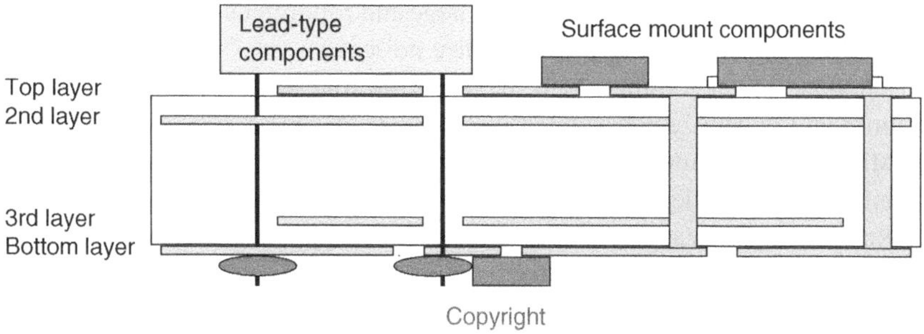

Figure 6.5 A four-layer board design

6.2.2 Component Placement

Component placement [2–5] is also critical in RF PCB design for several reasons. One of the primary reasons is signal integrity. The placement of components on the PCB affects the length and routing of traces, which in turn influences important factors such as impedance matching, signal propagation, and transmission line effects. By carefully considering component placement, it is possible to minimize signal degradation, crosstalk, and signal reflections, thereby improving the overall signal integrity of the RF circuit.

Another crucial aspect influenced by component placement is EMI/EMC considerations. RF circuits are particularly susceptible to both emitting and being affected by EMI. The placement of components on the PCB plays a significant role in the routing of high-frequency signals and the positioning of sensitive components. By strategically placing components, it becomes possible to reduce the coupling of noise and interference between components and mitigate the emission of unwanted RF energy, thus enhancing the EMI/EMC performance of the RF circuit.

RF performance is also greatly affected by component placement. Factors such as gain, noise figure, and linearity can be influenced by the placement of components on the PCB. For example, the distance between components can impact the coupling and isolation between RF stages. In an amplifier design, oscillation may occur if there is insufficient isolation between its input and output. By optimizing the placement of components, it is possible to minimize losses, improve gain and stability, reduce noise, and achieve better overall RF performance.

The design strategies are summarzied as [6]:

a) Identify and segment groups of circuits: Categorize circuits into groups such as antenna, analog, digital, switching, and audio.
b) Identify critical components: Determine the components that are crucial for the RF functionality and prioritize their placement.

c) Maximize grounding plane: Design a large and continuous grounding plane to ensure proper grounding and minimize noise.
d) Optimize power routings: Plan the power routing paths to minimize impedance and maximize power integrity.
e) Minimize traces and their lengths: Reduce the number and length of traces to minimize signal degradation and interference.
f) Rotate components with different angles: Rotate components to optimize their placement and reduce trace lengths.
g) Good I/O assignment: Assign input/output connections strategically to minimize trace lengths and optimize signal integrity.
h) Optimize PCB shapes or mounting holes: Design the PCB shape or include mounting holes to accommodate the layout and minimize interference.
i) Use daughter boards: Consider using daughter boards for specific circuits to simplify the layout and improve modularity.
j) Place components close to Integrated Circuits (ICs): Position components, especially RF, IF, and audio components, as close as possible to their corresponding ICs.
k) Group components with more interconnections: Cluster components that require frequent interconnections to minimize trace lengths and simplify routing.
l) Proper bus/ports assignment: Assign buses and ports intelligently to shorten trace lengths and avoid crossover.
m) Signal isolation: Keep the input and output of amplifier circuits physically separated to prevent signal coupling and potential oscillation.
n) Maintain distance between inductors/transformers: Avoid placing inductors and transformers too close to each other to minimize interference.
o) Orthogonal placement of neighboring inductors: Position neighboring inductors at right angles to reduce mutual inductance.
p) Good component placement eases routing effort: A well-optimized component placement layout simplifies the routing process and improves overall PCB design efficiency.

Figure 6.6 showcases a placement example of a 2.4 GHz transceiver module, where the components are strategically grouped from the antenna, matching elements, transceiver chip, to the interface with the host microprocessor (MCU). The transceiver chip is rotated by 45° to facilitate easier trace routing, considering the limited allowable width of the board.

6.2.3 Grounding Methods and Techniques

Grounding is the zero-voltage reference point of a circuit and also the current return path. Proper grounding is the prerequisite of good RF and EMC

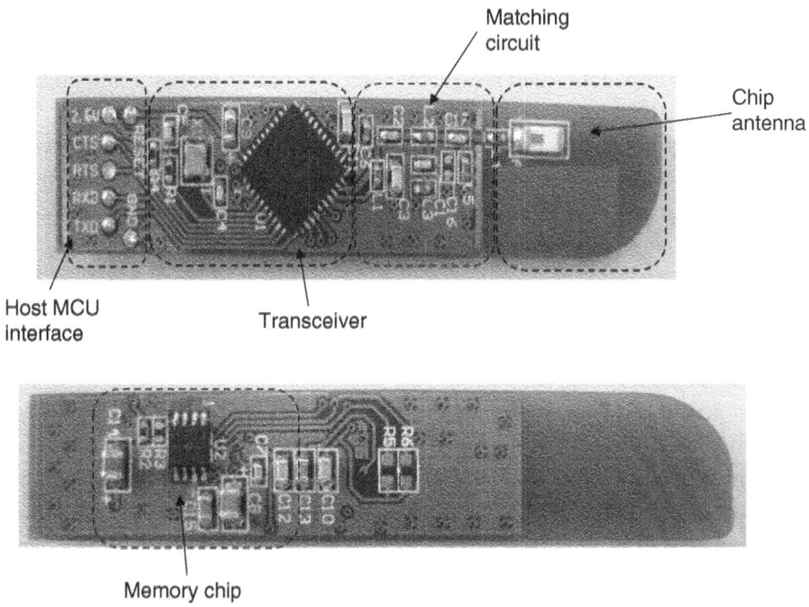

Figure 6.6 Component grouping on a 2.4 GHz transceiver module

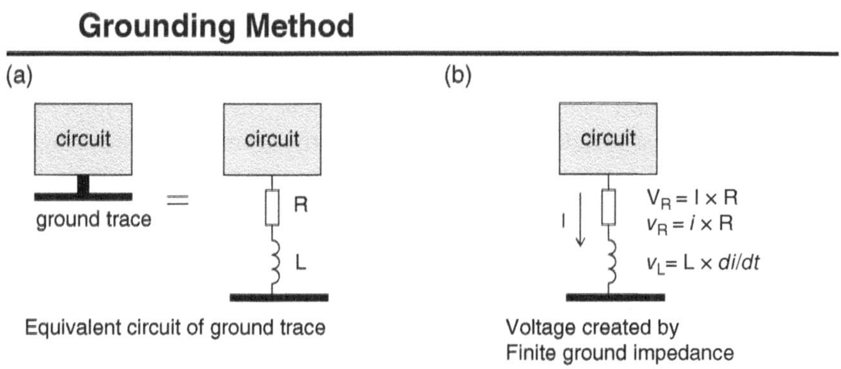

Figure 6.7 Finite ground impedance (a) equivalent ground connection and (b) current-induced ground voltage

performance [6–8]. The ground trace or ground plane can be modeled as an equivalent circuit of a resistor in series with an inductor as shown in Figure 6.7. If the ground trace is long and narrow, the ground resistance and inductance will be high. The return current will develop a voltage due to the finite impedance of the ground connection. If the circuit is an amplifier, the output signal can be coupled to the input due to finite ground impedance. Such feedback from output to input will cause unintended oscillation. On the other hand, if the circuit is an oscillator, the ground impedance will reduce the loop gain of the oscillator and present the circuit from oscillating.

The situation will become more complicated if there are multiple circuits with their grounds connecting to different parts of the main ground trace as shown in Figure 6.8. The ground current of each circuit may generate signal voltage and couple to other circuits causing interference among different building blocks. For RF designs, such interference or coupling results in poor block-to-block isolation. For receiver designs, poor receiving spurious response may be resulted as the RF filters may become ineffective. For transmitter design, the output harmonics will be very high as the harmonic suppression filter will also be ineffective.

As a best practice, it is recommended to minimize the length and maximize the width of the ground trace. For a ground plane, it is ideal to make it as large and continuous as possible. It is advisable to include an abundant number of ground via holes for effective ground connections. These measures help establish a low-impedance path for return currents, reducing the interference with desired signals.

In mixed-signal designs, separate ground planes can be employed for different circuits, such as digital, analog, and RF. It is crucial to ensure that all ground planes connect to the same endpoint, forming a star connection, as depicted in Figure 6.9. In cases where space or cost constraints exist, only one ground plane may be feasible. In such situations, a common, solid, and complete ground plane encompassing all circuits is necessary, as illustrated in Figure 6.10.

Overall, proper grounding is essential for achieving optimal performance, signal integrity, and EMC compliance in RF PCB designs. Therefore, it is important to follow these best practices for grounding. Firstly, grounding ensures signal integrity within RF systems. In RF circuits, maintaining signal integrity is critical. Grounding helps reduce noise, minimize signal reflections, and control

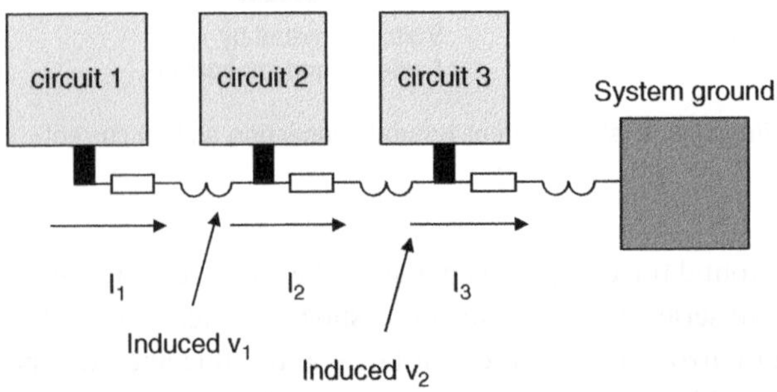

Figure 6.8 Interference between circuits by finite ground impedance

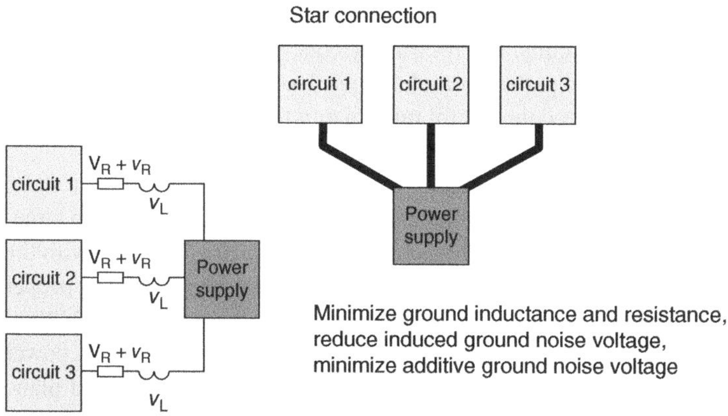

Figure 6.9 Star ground connection

Figure 6.10 Multipoint grounding connection

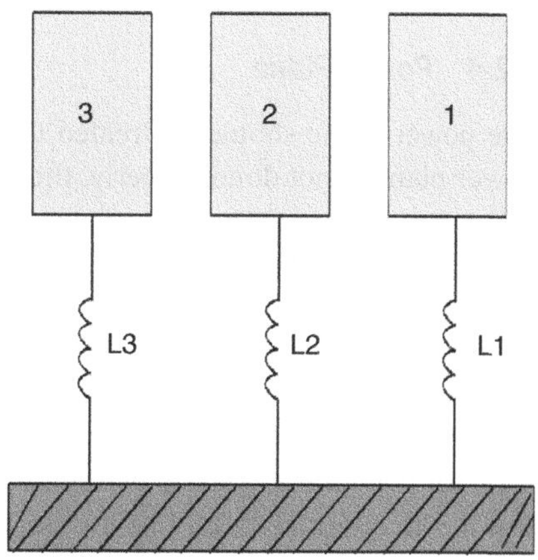

impedance. By providing a solid ground plane, it establishes a low-impedance path for return currents, preventing interference with the desired signals.

Secondly, grounding is essential for achieving EMC. RF circuits are sensitive to EMI from external sources and other circuitry on the PCB. Proper grounding shields RF circuitry from external EMI and prevents RF signals from radiating and interfering with other components. This helps maintain good isolation between RF components and other sections of the PCB. Thirdly, proper grounding ensures optimal performance of RF components. Components such as amplifiers, filters, and antennas often require a stable reference point or ground for proper operation. Inadequate grounding can lead to degraded performance, increased noise levels, and reduced sensitivity or gain of these components.

Furthermore, grounding facilitates the establishment of a clear return current path. In high-frequency circuits, the return currents associated with RF signals must have a low-impedance path to complete their loop. A well-designed ground plane provides this low-impedance path, minimizing the loop area and reducing the chances of electromagnetic coupling between traces.

Effective grounding also helps reduce ground loops. Ground loops occur when there are multiple ground connections with different potentials, resulting in circulating currents. Ground loops introduce unwanted noise and can interfere with the proper functioning of RF circuits. By ensuring a single-point ground and proper connection of all ground points, the risk of ground loops is minimized. Additionally, proper grounding aids in thermal dissipation [9]. RF circuits, especially power amplifiers or high-power components, generate heat. An adequate ground plane helps spread the heat across the PCB, improving thermal dissipation and preventing localized hotspots that can affect the performance and reliability of the RF circuitry.

6.2.4 Power Plane

The power plane should be treated the same as the ground plane [9, 10]. If the power plane is not done properly, the same issues of having a poor ground plane will occur. Furthermore, using "power planes" to distribute power across the PCB is considered good practice. Power wiring inductance and impedance to the components can be drastically reduced by using power planes. High-speed digital design can greatly benefit from this practice. The use of power planes is considered good design practice and can even be implemented on double-sided boards, especially when most of the signal tracks are on the top layer.

Power planes are typically placed in the middle layers of the board, usually closest to the outer surfaces. In the case of a four-layer board with complex power requirements, it is common to dedicate one layer to the ground plane and another layer to the various positive and negative power tracks. The ground rail, which serves as the signal reference line, is usually given priority over the power plane. Special Power Plane layers are often provided in PCB packages, and these layers are designed and laid out differently compared to the normal signal layers. In a normal signal layer, the board is assumed to be blank, and signal traces are added to become the actual copper traces. However, in a power plane, the board is assumed to be covered with copper, and laying down traces actually removes the copper. This concept may require some adjustment.

A simple power plane is a solid copper layer without any traces or removed copper areas. In such cases, there is no need to add traces to remove copper. However, for more complex boards, it is common to "split" the power plane by incorporating traces. This splitting is often done to separate the analog and digital grounds, reducing the coupling of digital ground noise into the more sensitive analog circuitry. A typical split power plane configuration involves placing a trace from the input power connector or main filter capacitors to the opposite edge of the board. Care should be taken to prevent unintentionally creating a power "loop" by accidentally connecting the two halves of the plane on the other side of the board.

6.2.5 Bypassing and Decoupling

The decoupling of power supplies is a vital aspect of RF PCB design for several important reasons. Firstly, RF circuits operate at high frequencies, which makes them highly susceptible to power supply noise or fluctuations. These disturbances can introduce unwanted interference and degrade the overall signal quality. By strategically placing decoupling capacitors near the RF components, a localized reservoir of charge is created. This helps stabilize the power supply voltage and minimize voltage fluctuations caused by rapid changes in current demand. As a result, a clean and stable power supply is ensured for the RF circuitry, reducing the likelihood of signal degradation.

In RF circuits, the reduction of noise is also crucial as they are vulnerable to various noise sources, including power supply noise. Decoupling capacitors act as bypasses for high-frequency noise, providing a low-impedance path for noise currents to flow away from the sensitive RF components. In doing so, they attenuate power supply noise and prevent it from affecting the performance of the RF circuitry.

Furthermore, stability and resonance control are critical considerations in RF circuit design, particularly when active components such as amplifiers and oscillators are involved. These components can be sensitive to power supply variations, which can potentially lead to instability or unwanted oscillations. Decoupling capacitors address this concern by providing a local energy source and damping any resonances that may arise due to parasitic inductances and capacitances in the power distribution network. This ensures stable and reliable operation of the RF circuit.

The location of the decoupling components [8, 10, 11] is critical. They should be placed in optimal locations, i.e., placing the RF one as close as possible to the chip. Each circuit block should be decoupled individually if there are multiple functional blocks on the PCB. For ICs with multiple power supply pins, each of them should have an individual decoupling capacitor since their pin locations may be far apart. Figure 6.11 illustrates the use of separate decoupling capacitors for different Vcc pins of the RF circuit of a 900 MHz transceiver chip, even though they are connected to the same supply voltage of 3 V. Multiple decoupling capacitors may be required for a mixed-signal PCB due to the frequency characteristics of the

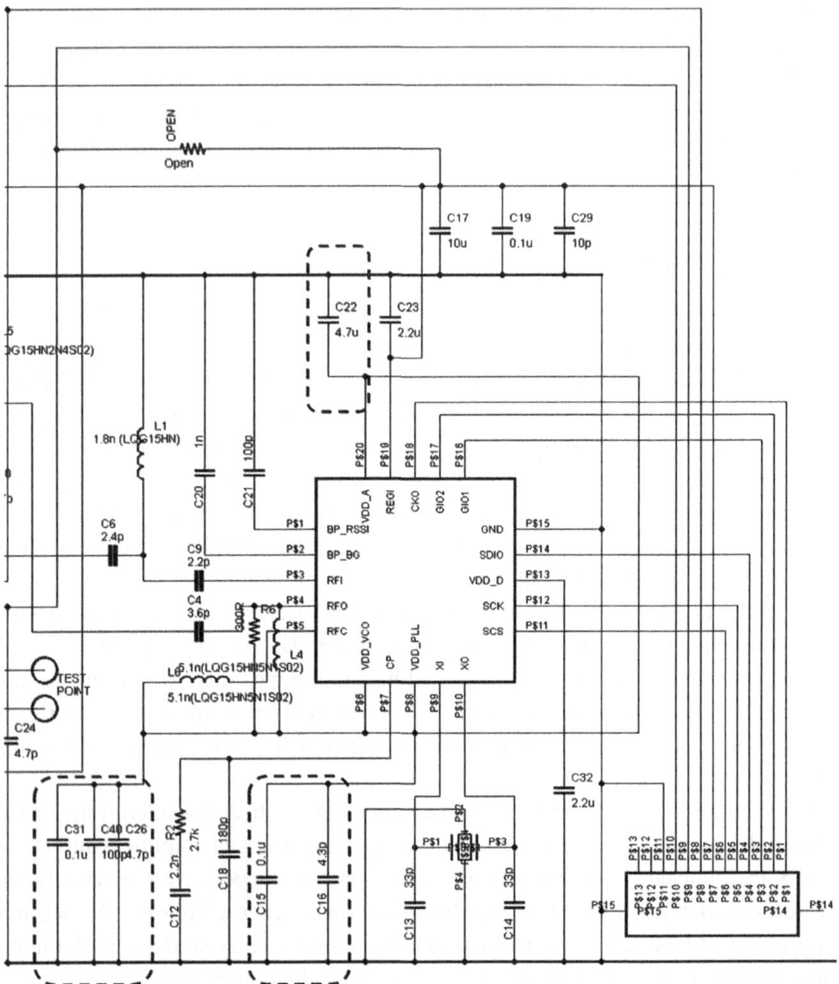

Figure 6.11 Multiple Vcc pins

capacitors resulting from their parasitic elements. Figure 6.12 shows the high-frequency model and impedance characteristics of a typical ceramic chip capacitor. The capacitor will behave as an inductor, making it ineffective when the operating frequency is beyond its self-resonance frequency. For the best decoupling result, it is recommended to use a decoupling capacitor with a self-resonance frequency that matches the frequency of the signal to be decoupled, as the reactive impedance of the capacitor is the lowest at that frequency. The resonant frequency of capacitors depends on their structure and value. In a mixed-signal PCB, it is

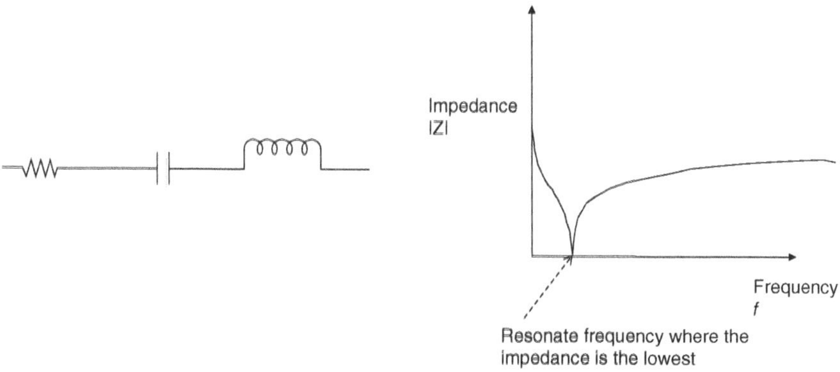

Figure 6.12 High-frequency model and impedance characteristic of a typical ceramic chip capacitor

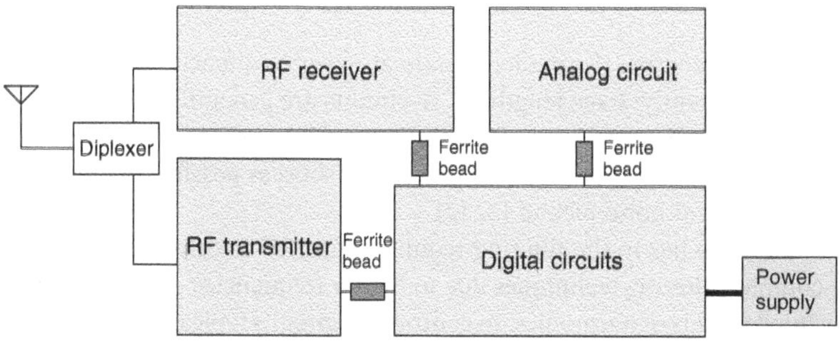

Figure 6.13 Using ferrite bead to isolation circuit from others

necessary to decouple the coexisting RF, IF, analog, and digital signals. Multiple decoupling capacitors in parallel with different self-resonant frequencies may be required to decouple all the signals, as shown in Figure 6.11. For high-performance devices or systems, a ferrite bead can be placed in series with the supply voltage of each subcircuit in addition to the decoupling capacitors, as shown in Figure 6.13. The impedance of a ferrite bead increases with frequency in a wide frequency range, isolating the concerned circuit from others due to the high impedance it presents.

To achieve optimal decoupling, it is recommended to utilize a decoupling capacitor with a self-resonance frequency that matches the frequency of the signal to be decoupled, as this minimizes the reactive properties of the capacitor. The resonant frequency of capacitors is determined by their structure and value. In the case of a mixed-signal PCB, where RF, IF, analog, and digital signals coexist, it is necessary to decouple all these signals. Figure 6.11 illustrates that multiple decoupling capacitors in parallel, each with different self-resonant frequencies, may be required to effectively decouple all the signals.

6.3 PCB Design for Other Circuits

6.3.1 IF Circuits

When designing a PCB for an IF circuit compared to an RF circuit, there are notable differences in the design considerations. One key difference is the voltage gain in IF amplifiers, which is typically much higher than in RF amplifiers. IF amplifiers can have gains of 60–100 dB in an RF receiver chip, whereas RF amplifiers typically have gains of 10–20 dB. This high gain in IF circuits requires special attention to component placement to minimize feedback from the output to the input. Without proper precautions, the IF circuit can easily oscillate due to its high gain and feedback.

In terms of trace lengths, IF circuits operate at lower frequencies compared to RF circuits. Consequently, trace lengths in IF circuits are generally not as critical as in RF circuits. This provides more flexibility in routing the signals on the PCB. However, it is still important to keep trace lengths as short as possible to minimize signal degradation and noise pickup [3, 12].

Another difference lies in the filtering requirements. RF circuits often necessitate more complex filtering techniques due to higher frequencies and the need to reject unwanted signals or harmonics. In contrast, IF circuits typically employ simpler filter designs that can be implemented using passive components such as resistors, capacitors, and inductors. In the case of ICs, IF filters can be easily integrated using either analog or digital approaches.

6.3.2 Baseband Circuits

Noise management is also important in baseband circuits. While they are generally less sensitive to noise compared to RF circuits, noise can still impact their performance. To mitigate noise, it is important to employ appropriate filtering techniques at the input and output stages to suppress unwanted noise sources. Proper grounding practices should be followed, including the separation of analog and digital ground planes, to minimize noise coupling between different circuit sections.

Crosstalk mitigation is crucial in baseband circuits, particularly when sensitive analog signals are in close proximity to digital signals. Maintaining appropriate spacing between traces carrying different signals, using shielding techniques such as guard traces or ground planes between sensitive signals, and employing proper termination techniques are effective methods for reducing crosstalk.

6.3.3 Audio Circuits

When designing audio circuits, maintaining signal integrity is of utmost importance in order to preserve the fidelity of the audio signals. This entails minimizing noise, distortion, and interference. To achieve this, it is crucial to keep trace lengths as short as possible and avoid routing audio traces near high-speed digital signals or noisy components. Additionally, implementing proper grounding techniques is essential to minimize the occurrence of ground loops and noise coupling.

Thoughtful component placement is another significant aspect to consider in audio circuit design. Placing sensitive audio components away from noisy components or high-frequency sources helps minimize interference. Careful attention should also be given to the physical layout of the PCB to reduce the length of analog audio traces and prevent them from being routed near areas of the board that may introduce noise.

Achieving low noise levels is critical for ensuring high-quality audio reproduction in audio circuits. This can be accomplished by utilizing low-noise components, designing amplifier stages with careful consideration to minimize noise, and employing appropriate shielding and filtering techniques to reduce the impact of external noise sources. Additionally, the use of balanced audio connections and differential signaling can aid in reducing noise pickup.

6.3.4 Power Supplies

Maintaining power integrity is of most importance in power supply circuits, especially in the case of switching regulator types that generate high-frequency switching noise. The objective is to ensure stable and clean power delivery by minimizing voltage drops, noise, and fluctuations. To achieve this, it is crucial to utilize wide and short traces for power connections, as this helps reduce resistance and inductance. Additionally, placing decoupling capacitors near power supply pins plays a significant role in filtering out high-frequency noise and stabilizing the voltage.

Effective thermal management is crucial in power supply circuits due to the heat generated, particularly in voltage regulators or power conversion components. To prevent overheating and ensure the reliability and longevity of the circuit, it is important to employ appropriate heat sinks and thermal vias and ensure proper airflow in the PCB layout to dissipate heat efficiently.

Power supply circuits, particularly switching regulator types, can be a notable source of EMI. Implementing proper filtering techniques such as ferrite beads or common-mode chokes is essential for suppressing EMI emissions. Adhering to applicable EMI/EMC standards and guidelines is crucial to ensure compliance and prevent interference with other circuitry.

In certain power supply circuits, voltage isolation may be necessary to ensure safety or separate different voltage domains. This can be achieved by utilizing appropriate isolation techniques such as optocouplers or transformers, which provide electrical separation between circuit sections. Additionally, ensuring larger physical separation between components and traces is important to minimize potential issues related to voltage isolation.

References

1 Khandpur, R.S. (2006). *Printed Circuit Boards: Design, Fabrication, Assembly and Testing*. McGraw-Hill.
2 David L. Jones, *PCB Design, Revision A*, 2004, https://alternatezone.com/electronics/pcbdesign.htm
3 Ruff, D. et al. (2018). *The Hitchiker's Guide to PCB Design*. EMA Design Automation.
4 Pattnayak, T. and Thanikachalam, G. (2018). "Antenna design and RF layout guidelines," Application Note AN91445, Rev. H, Cypress Semiconductor.
5 "Antennas for short range devices," (2018). Application Note APL10045, Silicon Labs.
6 Lau, H. (2020). "RF PCB course material," IEEE Boston Chapter Course.
7 Montrose, M.I. (1996). *Printed Circuit Board Design Techniques for EMC Compliance*. IEEE Press, Wiley.
8 Montrose, M.I. (1998). *EMC and the Printed Circuit Board: Design, Theory, and Layout Made Simple*, 1e. Wiley-IEEE Press.
9 "Making waves with RF: a PCB engineer's guide to RF design principles," (2018). EMA Design Automation.
10 Kowalewski, A. (2000). "Partitioning for RF design," *Printed Circuit Design Magazine*.
11 Burns, L. M. (2000). "RF & Microwave design techniques for PCBs," *Proceedings, PCB Design Conference West*.
12 Bailey, M. (2011). "General layout guidelines for RF and mixed-signal PCBs," Tutorial 5100, Maxim Integrated Products.

Part III

Software Design

7

Embedded Software Development

7.1 Introduction

The role of embedded software design is recognized as pivotal in the development of modern electronic devices, influencing a range of applications from consumer products to critical industrial systems. This chapter will explore the details of embedded systems, highlighting their unique characteristics and the design methodologies employed to create efficient and reliable software solutions. Embedded systems are defined as specialized computing systems that perform dedicated functions within larger mechanical, electrical, or Internet of Things (IoT) systems. Unlike general-purpose computers, these systems are optimized for specific tasks, often facing stringent constraints on resources such as processing power, memory, and energy consumption. As technology continues to advance, the demand for innovative embedded solutions is increasing, necessitating a thorough understanding of the design flow and architectural considerations that underpin software development in this domain.

The essential steps in the design flow of application software for embedded systems will be explored. The process will begin with the understanding of requirements that guide development, followed by the definition of system architecture. Various architectural styles, including event-driven and real-time architectures, will be examined to illustrate how specific application needs are addressed. Additionally, a practical example of software architecture design will be provided, focusing on a tracking watch developed for construction worker safety and productivity. This example will guide readers through the process of defining subsystems, designing hardware, and establishing interfacing and timing requirements.

Finally, the critical differences between state machines and real-time operating systems (RTOSs) will be discussed, along with the factors that influence the selection of microcontrollers (MCUs) or System-on-Chips (SoCs) for embedded

Introduction to Wireless System Design: From Circuits to Web-based Applications, First Edition.
Henry Lau, Ludy Liu, and Keith Chan.

applications. By the end of this chapter, a comprehensive understanding of embedded software design principles and practices that pave the way for successful system implementation will be achieved.

7.2 Embedded System and Devices

An embedded device is designed to run software within a computer system, known as an embedded system, to perform dedicated tasks. An embedded system may consist of hardware, software, and mechanical components. Examples of embedded systems include cellular phones, wearable devices, medical devices, wireless sensors, security alarm systems, and safety-critical applications such as Advanced Driving Assistance Systems and Autonomous Driving (ADAS) [1]. The software running on an embedded system is referred to as embedded software, which is often required to perform dedicated tasks with strict timing requirements and limited Central Processing Unit (CPU) [2, 3] resources and memory.

An embedded system is a computer system that contains microprocessors, microcontroller units (MCUs), digital signal processors (DSPs), or SoCs, as well as Random Access Memory (RAM), Read-Only Memory (ROM), Flash memory, and input/output peripherals [2, 3]. The basic processing of a simple embedded system typically consists of an input processing task, a main loop state machine, and output processing [4]. The input data often originates from a hardware or sensor device. The most basic embedded system includes a main loop state machine that controls the primary flow of the system. This state machine continuously performs specific services or tasks, with state changes triggered by events (e.g., arrival of data) or driven by timing. Upon receiving input data, the main loop state machine will either change state or remain in the same state based on the input. After processing the input data, the application program computes the output data using a specific algorithm. The output data is then sent in a predefined format to a component within the same system or to an external system via on-board peripherals such as Inter-Integrated Circuit (I2C), Universal Asynchronous Receiver-Transmitter (UART), and Serial Peripheral Interface (SPI) devices [2, 3].

More complex embedded systems are required to operate within a domain-specific architecture that supports multiple cores, advanced memory technologies, and application-specific accelerators and peripherals. The features and characteristics of several notable architectural styles widely deployed in modern embedded systems will be discussed later in this chapter. Readers will gain insights into selecting a suitable architecture for their applications. Embedded systems often run with the support of a RTOS [4] to perform tasks with strict timing requirements, which will be discussed in more detail later in this chapter.

Embedded systems typically exhibit properties such as low cost, limited CPU resources, small footprints, and reduced power consumption compared to mainframe computer-based systems. To create a well-designed system, designers must start with a clear understanding of the requirements before moving to the design process. The design flow process of an embedded system will be discussed in the next section, guiding readers step by step in designing an efficient embedded system.

7.3 Design Flow of Application Software of Embedded Systems

Figure 7.1 shows the key steps involved in embedded product development. The embedded software development flow consists of six key steps. The first step is to define the system requirements. The system design must be based on requirements defined by the customer, product owners, and stakeholders. After the system architecture definition is complete, detailed design of the software architecture can follow based on the system requirements. To enable team members to collaborate on design ideas, there should be a well-written document that describes the details of the software architecture. The detailed software design should not begin until the software architecture definition has been completed. The software testing should be efficient and based on system requirements. Two common test methods, Black Box Testing and White Box Testing, will be discussed in the next chapter. The user interface of the system should be well-defined to enable efficient testing and seamless integration.

7.3.1 Understand the Requirements

In the first step of the software development flow, the system requirement specification phase, project stakeholders collaborate to define the services and features that the embedded system will provide. This group includes customers, product owners, architects, programmers, and engineers. It is essential to have a comprehensive understanding of the requirements to ensure that the resulting design aligns with the intended functionality. The specific intricacies of product development and project management processes are not covered in depth, as these processes are often influenced by local laws, company internal policies, and industrial standards that can vary widely.

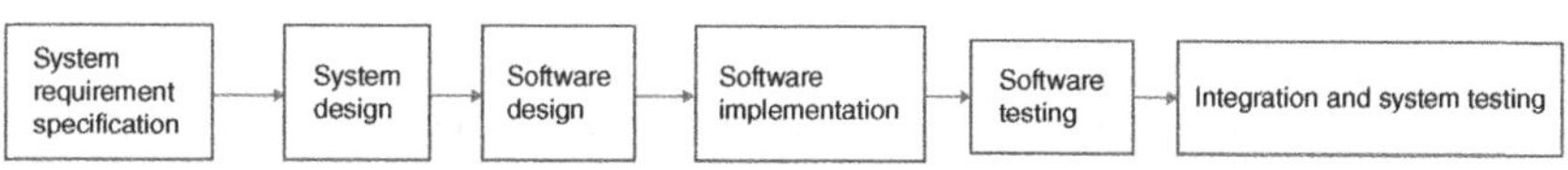

Figure 7.1 Embedded product development flow

The emphasis in this discussion is placed on providing guidance for architecture design based on the system requirements. These system requirements are frequently shaped by a variety of factors, including market demands, financial objectives, stakeholder needs, and design goals associated with the product under development and its subsequent launch. This section will examine the distinct characteristics and noteworthy features of several popular architecture styles that can be employed to meet the defined system requirements.

7.3.2 System Architecture Definition

The System Architecture design involves creating a comprehensive description of the software's overall structure and framework within a system. In order to collaborate with team members to address all the feature concerns and share ideas, the first step is to create a high-level system view which includes the interactions between system components. This is achieved through a high-level system diagram.

The system should be broken down into subsystems in order to ensure the design is modular and maintainable. Each subsystem should have a detailed description of its input, expected behavior, and output. In addition, providing state machine, sequence, and data flow diagrams helps to illustrate the behavior of the system. The interfaces between subsystems should be clearly defined to facilitate seamless communication and integration.

7.3.2.1 Choice of Architecture Style

The choice of architecture style depends on factors such as the system's requirements, constraints, selection of MCU or SoCs, hardware capabilities, real-time considerations, and the desired level of modularity and scalability. The following sections describe several common architectures in detail. The characteristics and benefits of using the architecture style will be explained to help readers make smart decisions when choosing a suitable style for their design.

7.3.2.2 Event-Driven Architecture

Event-Driven Architecture [4] supports the fast response of events originating from internal subsystems and external systems as shown in Figure 7.2. Both event-based and asynchronous communications are supported in this architecture style. State machines are used to handle the communications among subsystems. Events are triggered by state machine states or arrival of input data. Event-Driven Architecture supports fast event response from internal and external systems and utilizes state machines for communication among subsystems.

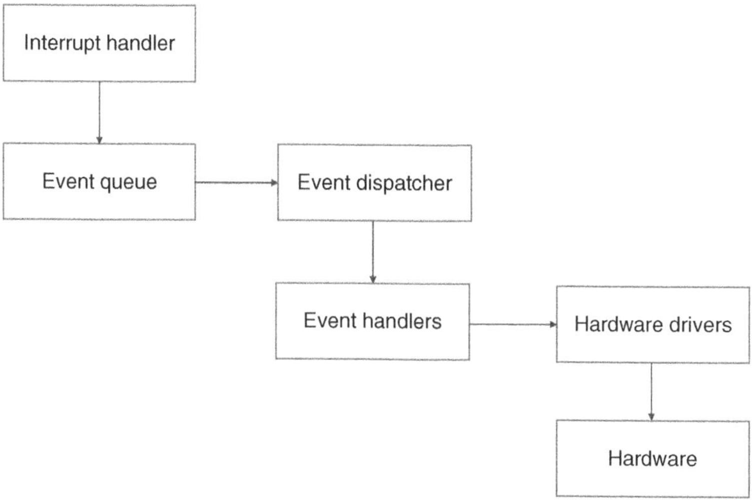

Figure 7.2 Block diagram of an event-based architecture for embedded software design

7.3.2.3 Component-Based Architecture

A Component-Based Architecture [5] is constructed by breaking down a system into subsystems, as shown in Figure 7.3. This architectural style is often used in embedded systems. Communication is accomplished through data exchanges

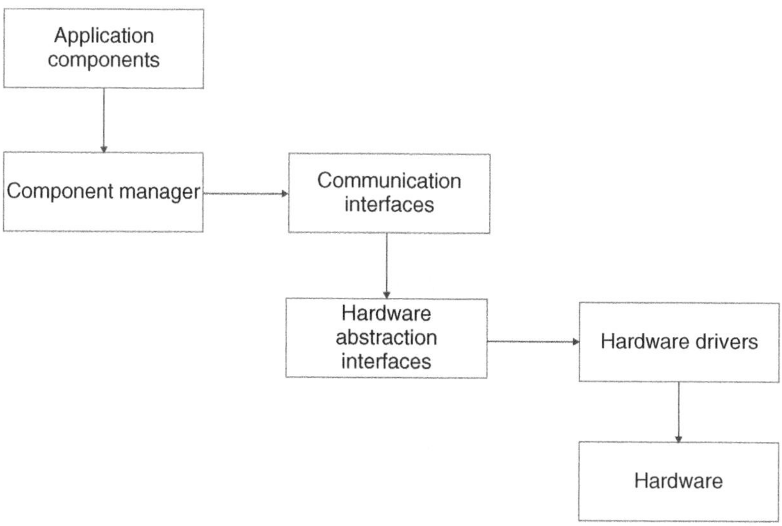

Figure 7.3 Block diagram of a component-based architecture for embedded software design

via a set of well-defined interfaces. This approach allows for the reuse of individual subsystems in case of changes to the MCU or hardware components. For example, in a high-end printer and scanner application, a subsystem is designed to scan the original color poster and output RGB data in a specific format. The output RGB data is then fed into a printing component in real time within specific timeframes. To support the large amount of RGB data coming into the printing component, it must have fast memory read and write access to fulfill the system's timing requirements. In this application, components with critical timing needs can be realized with a real-time-based architecture.

7.3.2.4 Real-Time Architecture

Real-Time Architecture [6] is used for systems or subsystems with specific timing requirements. This architecture should be utilized when specific tasks must be completed within particular time constraints [7]. An RTOS is often used to support the configuration of time-driven functionalities. The RTOS is also used for configuration of interrupt handlers, resource utilization, device drivers, memory management, and task priorities to support the timing requirements and deterministic behavior of the system [8]. One example of a real-time architecture is the collision avoidance system that is widely deployed in modern vehicles. The system uses sensors to keep track of vehicle speed, moving and stationary objects in the surroundings, and performs calculations to determine the possibility of potential collisions. In case the objects tracked in real time are getting too close, it provides visual and audible warnings to alert the driver to take action to avoid possible collisions. For simple embedded systems, an RTOS may not be needed. Figure 7.4 shows a real-time architecture that allows multiple tasks to be managed based on priority and timing constrains.

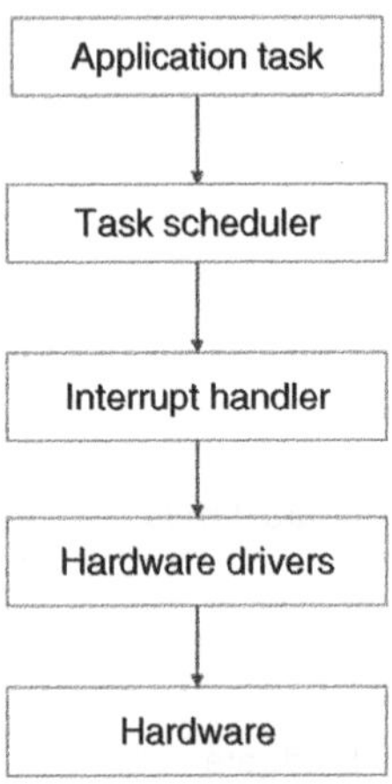

Figure 7.4 Block diagram of a real-time architecture for embedded software design

7.3.2.5 Hierarchical Architecture

Hierarchical Architecture [9] is often used in complex embedded systems to support industry standards, including security and safety standards. In this style, the system is divided into hierarchical layers based on abstraction levels or functionality, as shown in Figure 7.5. Each layer provides services or functionalities to adjacent layers, allowing for modularity and reusability [9]. A Wireless IoT Application is a good example illustrating such a hierarchical architecture. It can be structured into the following four layers: the Wireless IoT Devices layer, the Edge Devices or Gateways layer, the Network

Infrastructure layer, and the Cloud or Central Server layer [10, 11].

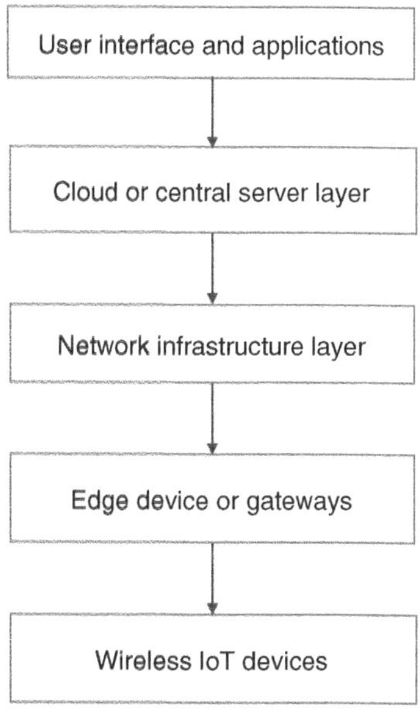

Figure 7.5 Block diagram of a hierarchical architecture for embedded software design

Wireless IoT Devices The lowest level consists of the wireless IoT devices themselves, such as sensors, actuators, and other connected devices. These devices have wireless communication capabilities to transmit and receive data wirelessly. The localized processing improves efficiency, reduces latency, and enhances responsiveness.

Edge Devices or Gateways At the subsequent level, edge devices or gateways act as intermediaries connecting wireless IoT devices to higher-level systems. These devices facilitate preprocessing, aggregating, and filtering of data originating from multiple wireless IoT devices before transmitting it to the cloud or central server. Security services are also supported in this layer.

Network Infrastructure Layer The network infrastructure layer provides the wireless connectivity capability that enables communication between the wireless IoT devices, edge devices, and other components of the architecture. The devices support on-board technologies, including cellular networks, Wi-Fi, Bluetooth, ZigBee, and others.

Cloud or Central Server Layer The cloud or central server provides a centralized and scalable platform for managing the wireless IoT data. It is responsible for storing and processing the data sent from the edge devices or gateways. It provides support for over-the-air software installation, data management, and data analytics. For AI- and machine learning-based applications, this layer also provides model training and algorithm optimizations.

User Interface and Applications At the topmost level, there are user interfaces and applications that allow users to interact with the wireless IoT system. These interfaces can be mobile applications, web portals, or other software applications. They provide support for monitoring and controlling the wireless IoT sensors connected to the network, as well as data analysis.

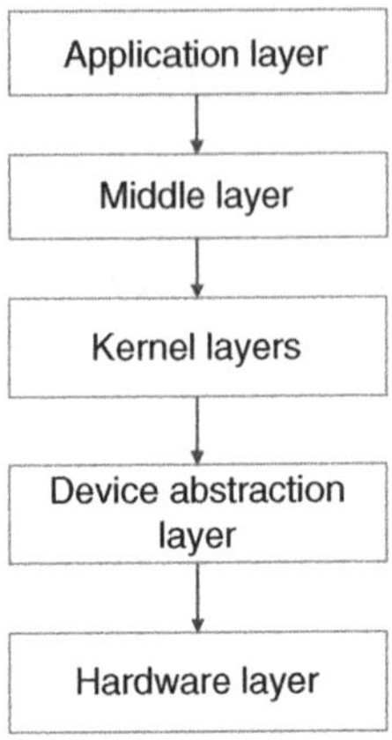

Figure 7.6 Block diagram of a layered architecture for embedded software design

For wireless IoT applications that require real-time operation, the hierarchical architecture is commonly used. This architecture also allows for scalable data storage, advanced analytics, and seamless integration with user interfaces and applications, providing a comprehensive and interconnected system for mobile IoT devices.

7.3.2.6 Layered Architecture

As shown in Figure 7.6, a layered architecture [12] is widely used in today's modern embedded systems that have critical safety requirements such as critical infrastructure monitoring, autonomous driving systems, and wearable medical devices. This architecture typically consists of several distinct layers. Each layer provides a level of abstraction to serve a specific purpose. The upper layers are built upon the services and abstraction supported by the lower layers. This allows modularity and reusability. In Layered Architecture, the Hardware Layer interacts directly with physical hardware components, while the Device Abstraction Layer provides an abstraction of the hardware layer through standardized interfaces and device drivers. The Kernel Layer supports operating system functionalities, and the Middleware Layer facilitates communication between system components. The topmost layer is the Application Layer, which implements application-specific functions and features.

Hardware Layer The first layer is the Hardware Layer that represents the physical hardware subsystem of the embedded system. The Hardware Layer interacts directly with the hardware components of the systems. This layer is responsible for the low-level interactions with the physical hardware. It handles all the data and signal communication of the hardware components. It also manages the communications of internal and external interfaces. It is possible for layers above the Device Abstraction Layer to directly communicate with the hardware layer based on system requirements. For example, if the hardware register can be accessed through an IO mapped address, the hardware registers can be read and written directly from the top application layer. One use case may be that the entire system needs to be reset due to triggered system faults. For example, suppose there is a drop in the power level of the system, and the system needs to be put in sleep mode without delay for safety reasons. The top layer may have access to the application programming interface (API) calls that can be used to immediately reset the hardware without risking potential blocks from the operating system.

Device Abstraction Layer The second layer is the Device Abstraction Layer. This layer provides an abstraction of the underlying hardware layer. The advantage of having a layer with device abstraction is to isolate the support of device drivers that handles all the communication among hardware components in one single layer. Communication between the device abstraction layer and the hardware layer is accomplished through a set of well-defined standard API library calls. The standard APIs allow the specific implementation of hardware to be hidden. This allows modularity and adaptable hardware and system integration. When a MCU, SoC, or DSP license is purchased, the software libraries for specific device drivers of all on-board peripherals are often supported by the hardware providers and/or approved third party Board Support Package providers. This allows minimal effort of developing device drivers and the development team to focus on software development of the system features and functionalities. In today's market, development teams may need to develop their software on prototype hardware such as device drivers. It is important to know that it is possible that some device drivers may not work as expected and debugging these device drivers may become a challenge. These kinds of issues may occur during the integration of new hardware. The verification of device drivers are usually performed during hardware testing.

Kernel Layers The third layer is the Kernel Layer, also known as the Operating System Layer. The Kernel Layer provides support for the operating system functionalities including timing control, timer access, memory management and task management. Besides providing the common operating system functionalities mentioned above, the kernel layer can also provide API library calls to support CPU bandwidth management in the system. In some OS such as QNX OS, specific tasks can also be configured to consume specific CPU bandwidth. For example, a critical process running on the target processor can be configured to use up to 60% of the total available bandwidth. The Kernel can also provide service to measure memory consumption and the CPU bandwidth consumption among all cores. For example, the "Top" command used in the Linux OS and QNX OS. The kernel layer is hardware dependent.

Middle Layer The Middleware Layer, also known as the Real-Time Environment Layer, sits above the Kernel Layer. This layer provides services and functionality to support communication between system components. In many modern embedded systems, an SoC offers support for multiple general-purpose processor cores and a few application-specific DSP cores. The main processor is responsible for running the primary state machine of the embedded system. On the other hand, the application-specific DSP may handle critical tasks that require fast response times, supported by floating-point mathematics and hardware accelerators.

Data transfer between the DSP and general-purpose processors occurs via inner-core communication. Both the main processor and DSP have access to shared memory. In such systems, it is beneficial for the Middleware Layer to provide a service for transmitting data from the DSP to the main processor. The size of memory allocated to shared memory may not change, but the physical addresses are usually hardware-dependent. The communication method for inner-core communications can remain unchanged even if there are modifications to the DSP hardware.

The DSP cores are often assigned to run specific calculations, and the output data may need to be sent to an external interface. However, the peripherals used to transmit data to an external system are generally available only on the main processor side. In this case, the data must be copied from the DSP cores to the general processor through inner communications. This layer often includes communication protocols, inter-process communication between different cores, and support for specific peripheral device drivers.

Application Layer The topmost layer is the Application Layer. This layer implements the state machine of the application program to support specific functions and features of the embedded system. It often provides support for the user interface, debugging interface, control algorithms, memory storage, and interfaces for external communications. Direct communication between the application and kernel layers is permitted depending on the specific requirements of the system. For example, the Application Layer may need to calculate the required runtime to complete the main loop of the application software. Instead of using the API call in the Middleware Layer, it can directly read the I/O mapped address of the timer register through an API call provided by the Kernel Layer.

7.4 Software Architecture Design Example

7.4.1 High-Level Block Diagram

The first step for software architecture design is to draw the high-level block diagram to show all the required systems and components to meet the system requirements. A tracking watch for construction workers will be used for this section to illustrate all the steps for designing the software architecture. Figure 7.7 is a high-level block diagram of a Tracking Watch for Construction Workers, showing all the components to be supported in the system. It includes both hardware and software components.

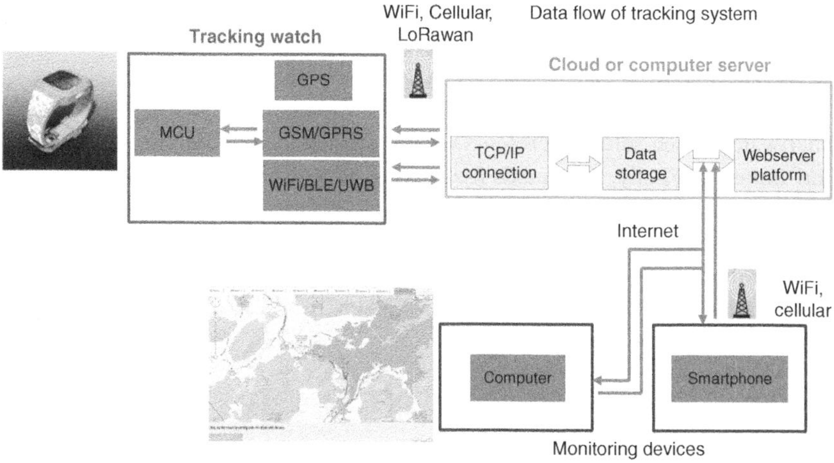

Figure 7.7 High-level system block diagram [Courtesy of Lexiwave Technology (Hong Kong) Limited]

7.4.2 Tracking Watch for Construction Worker Safety and Productivity

The construction industry is inherently hazardous, with workers facing a wide array of risks on job sites, ranging from heavy machinery operations and falling objects to exposure to harsh environmental conditions. In this dynamic and demanding environment, the need for effective worker tracking and monitoring solutions has become increasingly paramount. By implementing a comprehensive tracking system, construction companies can not only enhance the safety and well-being of their workforce but also drive improvements in overall productivity and job site management. One of the primary advantages of a tracking system for construction workers is the ability to monitor their real-time location and movements. GPS-enabled tracking devices can provide precise geolocation data, allowing supervisors to keep a watchful eye on the whereabouts of their workers, and ensure they are operating within designated safe zones. This geofencing capability can trigger alerts when workers venture into potentially hazardous areas, enabling a swift response to mitigate risks and prevent accidents.

Beyond location tracking, a well-designed tracking solution should also incorporate advanced sensors to monitor a range of environmental and physiological factors. Integrating sensors to measure ambient temperature, humidity, and air quality can help identify and address hazardous working conditions, protecting workers from heat-related illnesses, respiratory problems, and other environmental health risks. Furthermore, the inclusion of biometric monitoring, such as heart rate and blood pressure tracking, can provide valuable insights into the worker's physical well-being, enabling early intervention, and preventive measures to safeguard their health.

While handheld devices can offer some tracking functionalities, a wearable smartwatch-based solution presents several distinct advantages for the construction industry. Firstly, the hands-free nature of a wrist-mounted device allows workers to maintain full focus on their tasks, enhancing their situational awarenes, and reducing the risk of accidents. Additionally, the rugged and water-resistant design of a construction-grade smartwatch is better suited to withstand the demanding job site environment, minimizing the risk of damage, or loss compared to a fragile handheld device.

7.4.3 Defining the Subsystems

To implement a complete Construction Worker Safety and Productivity system, in addition to the tracking watch, several key components and systems would be required. First, a Central Monitoring and Data Management Platform is crucial. This would be a centralized web-based or cloud-based platform to receive, process, and analyze the data collected from the tracking watches. The platform should provide a user-friendly interface for supervisors and managers to monitor worker activities, receive alerts, and generate reports. It should also offer data visualization tools, historical trend analysis, and integration with other job site management systems.

A Mobile Application would also be necessary as a companion tool for workers to interface with the tracking system. The app should allow workers to view their own data, receive alerts and notifications, and communicate with the central monitoring system. It can also serve as a channel for two-way communication, enabling workers to report incidents, request assistance, or provide feedback.

The system would require a reliable Communication Infrastructure to enable real-time data transmission between the tracking watches and the central monitoring platform. This could involve a combination of Wi-Fi, cellular, or proprietary wireless protocols (e.g., LoRaWAN, Zigbee) depending on the job site's infrastructure and coverage requirements. The communication network should provide seamless connectivity and ensure secure data transmission. Sensor Integration is another critical component, involving the integration of additional environmental and safety sensors at the job site to complement the data collected by the tracking watches. Examples include air quality and toxic gas sensors, RFID-based proximity sensors, vibration sensors, heart rate and blood pressure sensor, as well as temperature and humidity sensor.

As mentioned in the previous section, this tracking watch is a good example illustrating such a hierarchical architecture of Wireless IoT Applications, structured into the following four layers: the Wireless IoT devices layer, the network infrastructure layer, the cloud or central server layer, and the smartphone application software. The central server and smartphone application software designs will be discussed in the next few Chapters.

7.4.4 Designing the Hardware of the Tracking Watch

With the system requirements and necessary hardware identified, it is now time to design the actual watch. Previous sections have outlined the desired features and required subcomponents. Designing a tracking watch is particularly challenging compared to other consumer products due to its small size and limited battery capacity. Compromises will need to be made to include the most critical and desirable features. Figure 7.8 shows a block diagram of a practical watch design that incorporates all the essential features and required components.

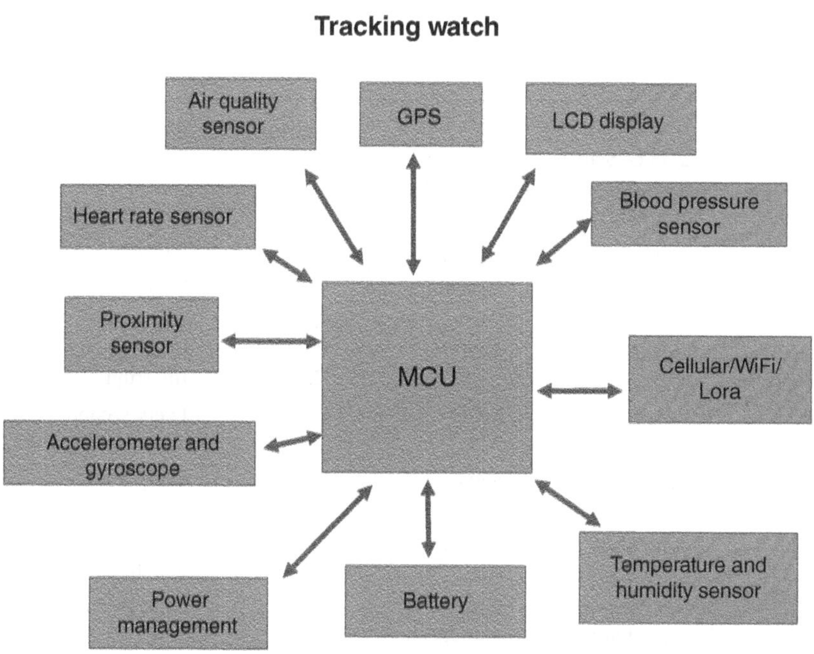

Figure 7.8 Block diagram of a practical tracking watch

7.4.5 Interfacing Requirements

To optimize the software design for the tracking system, it is crucial to determine the interface requirements, including the hardware needs, timing, and data flow between the watch's core, the MCU, and its peripherals. Table 7.1 displays the interface requirements between all the components of the tracking system, including the server system, to provide a comprehensive picture of the data flow. Focusing on the embedded software design for the watch, the core of the watch is the MCU, which interfaces with all the sensors, LCD display, Power Management Chip, GPS, and Wireless Modules. Therefore, it is essential to select a suitable MCU with sufficient Input/Output (I/O) pins and communication ports to support the required functionalities.

MCUs typically come in a range of models with varying RAM, Flash memory sizes, clock speeds, and pin counts. As the number of required pins increases, the physical size of the MCU also grows. To optimize the design, a trade-off and optimization process would be necessary to minimize the number of pins while still providing the required functionalities. This can be accomplished by carefully selecting the sensors and modules. It would be advantageous to use sensors that communicate using I2C or SPI interfaces, as these communication protocols allow for parallel connections and do not consume additional pins from the

Table 7.1 Interface requirements for the tracking watch

Data Source	Data to be Sent	Data Destination	Interface Method	Data Exchange Frequency
Watch	Sensor data, location	Server	Wireless	Every 1–5 s
Server	Sensor data, location	Smartphone	Wireless	Every 1–5 s
Server	Sensor data, location	User computer	Wireless or internet	Every 1–5 s
Various sensors	Sensor data	MCU	I2C, SPI, UART	Every second or by Interrupt
GPS module	Location	MCU	UART	Every second
Wireless module	MCU	Server	UART	Every 5 s
MCU	Graphic data	Display	SPI, parallel I/O	Only when updating display
MCU	Power control/ battery charging	Power management chip	I/O	Once in a while

MCU. On the other hand, many wireless modules use the UART interface, which does not support parallel connections. In such cases, it is crucial to ensure that the selected MCU has a sufficient number of UART ports to accommodate the required wireless modules.

7.4.6 Timing Requirements

The communication timing requirements between the MCU and its various interfacing components are crucial for the efficient operation of the tracking system. The watch must be able to transmit sensor data and location information to the server at a frequency of every 1–5 seconds. This means the MCU must send this data to the wireless module every 1–5 seconds to ensure the server receives the latest information in a timely manner. Similarly, the server must be able to transmit the sensor data and location information to the smartphone or user computer at the same frequency as the watch-to-server communication, allowing the end user to access the most up-to-date tracking data. This requires the server to receive the data from the watch and then forward it to the client devices without any significant delays.

Regarding the sensors, the MCU must be able to read the sensor data either at least once per second or through interrupt-driven communication, depending on the sensor interface (I2C, SPI, or UART). This ensures the MCU has the latest sensor data available for transmission to the server. The GPS module, which provides the location data, must also communicate with the MCU at a rate of once per second to ensure the transmitted location information is current. The MCU must also send the sensor data and location information to the wireless module every 5 seconds, matching the frequency of the watch-to-server communication. This allows the server to receive the latest tracking data in a timely manner. For the display, the MCU only needs to update the graphic data when necessary, as the display does not require constant updates. Similarly, the communication between the MCU and the Power Management Chip for power control and battery charging can be done on an "as-needed" basis, without the need for high-frequency updates.

7.4.7 Software Design Strategies

Given that the MCU is expected to spend most of its time processing communication tasks, there are several strategies that can be implemented to enhance its processing capabilities. One key strategy is to implement a priority-based task scheduling system [4, 13] within the MCU firmware. This ensures that high-priority communication tasks, such as sensor data and location updates, are executed in a timely manner. Lower-priority tasks, like display updates or power management, can be scheduled during the MCU's idle time or using a lower-priority interrupt.

Another important aspect is interrupt optimization. By carefully designing the interrupt handling routines to minimize the processing time and reduce the impact on the MCU's overall performance, the communication tasks can be handled more efficiently. Utilizing efficient interrupt service routines [4, 13] that only perform the necessary tasks and return as quickly as possible can significantly improve the MCU's responsiveness. Offloading communication tasks to the peripheral devices themselves can also be beneficial. For example, by using the Direct Memory Access capabilities of the MCU, data can be transferred between the peripheral interfaces and the memory without the need for constant MCU intervention, freeing up the MCU for other tasks.

As mentioned previously, implementing a multiplexing scheme to share communication interfaces (I2C, SPI) between multiple sensors and peripherals can also optimize the communication throughput and reduce the number of MCU pins required.

Firmware optimization techniques, such as reducing unnecessary memory allocations and memory copies, utilizing look-up tables or other data structures to minimize computational complexity, and employing loop unrolling and function inlining, can significantly improve the efficiency of the MCU's code.

For certain noncritical communication tasks, such as display updates or power management, considering the use of asynchronous communication models can allow the MCU to handle these tasks in the background without interrupting the high-priority sensor and location data exchanges.

Depending on the required features of the watch, the MCU can run in a state machine [14] or an RTOS. Integrating an RTOS into the MCU firmware can provide a structured task management and scheduling framework, simplifying the handling of multiple concurrent communication tasks and improving the overall system responsiveness.

7.5 State Machine Versus Real-Time Operating System

Choosing between a state machine and an RTOS for real-time applications requires careful consideration of their respective advantages and disadvantages. A state machine is a model of computation that represents the different states an application can be in, along with the transitions between those states. It offers a straightforward and intuitive way to design and implement system behavior. In contrast, an RTOS is a specialized operating system designed for deterministic and predictable real-time performance, supporting preemptive multitasking, efficient interrupt handling, and comprehensive resource management.

When evaluating the choice between a state machine and an RTOS for a real-time application, several factors must be considered. A state machine provides well-defined states and transitions, leading to predictable behavior and making the code more readable and maintainable. It optimizes resource use by executing only the necessary code for the current state, and its modular design facilitates easier reuse and scalability. Additionally, a state machine simplifies error handling and recovery mechanisms. However, it may struggle with highly concurrent or parallel tasks, as it is better suited for sequential and event-driven applications. As complexity grows, maintaining and debugging a state machine can become challenging. Furthermore, it typically lacks preemptive scheduling capabilities and advanced resource management features found in an RTOS, such as dynamic memory allocation and task scheduling.

In contrast, an RTOS ensures deterministic performance through preemptive multitasking, allowing the system to interrupt and switch between tasks based on priority. It also offers efficient interrupt handling and comprehensive resource management features, which enhance the overall reliability and fault tolerance of the system. Using the tracking watch as an example, its main task is to collect measurement data from various sensors. The sensors and the MCU typically enter a power-saving sleep mode. The sensors will return to normal operating mode when the sensing parameter is below or above a preset threshold, providing an interrupt signal to wake the MCU to process the critical information. A state machine can handle these processes without difficulty. However, if the watch requires more user interaction and data exchange between the MCU, sensors, and the server, an RTOS would be more appropriate.

7.6 Selection of MCU or SoC

In the design of a tracking watch, selecting the right MCU is crucial to manage diverse sensor inputs, wireless connectivity, and processing requirements. Suitable options for this type of device include the STM Arm Cortex-M3 or M4 [15], the Nordic Semiconductor nRF52 or nRF53 Series [16], and the Texas Instruments MSP432 Series [17]. If the tracking watch evolves into a smartwatch with features akin to a smartphone, such as a high-quality camera, voice communication, and comprehensive wireless connectivity, a SoC would be necessary. The key differences between SoCs and MCUs lie in their integration level, complexity, and target applications. SoCs are highly integrated chips that combine a CPU, GPU, memory, and various peripherals on a single piece of silicon. This integration provides advanced computational capabilities, including graphics processing, multimedia support, and robust connectivity options. SoCs are typically found in sophisticated devices like smartphones, tablets, and high-end embedded systems, where powerful functionality is essential. In contrast, MCUs are generally less complex, integrating a processor, memory, and basic peripherals such as timers and communication interfaces. Although MCUs may lack the advanced features of SoCs, they are designed for power efficiency and cost-effectiveness, making them ideal for simpler embedded applications, IoT devices, and industrial control systems.

The design approach also differs between SoCs and MCUs. SoC design involves a comprehensive system-level approach, integrating multiple IP blocks to optimize overall performance. In contrast, MCU design focuses more on the microcontroller core and its peripherals, resulting in simpler system integration. Power consumption is another distinguishing factor. SoCs typically have higher power requirements due to their complexity and advanced features, while MCUs are engineered for energy efficiency, making them suitable for battery-powered or energy-constrained applications. SoCs often include external interfaces to support standard communication protocols such as USB, SPI, I2C, HDMI, UART, and Ethernet. For example, Figure 7.9 illustrates the architecture of an SoC system featuring two domain-specific processors for video image processing applications: one ARM processor for general-purpose tasks and a DSP for complex algorithms. Other domain-specific SoCs may be optimized for radar signal processing, network processing, or graphics processing, often supporting multiple cores, DSPs, and application-specific hardware accelerators for embedded systems.

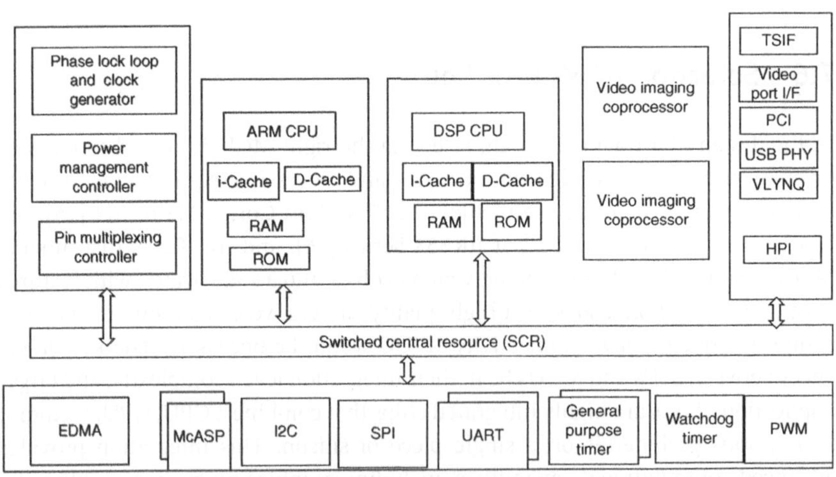

Figure 7.9 SOC with domain specific processors

References

1 Choi, N. et al. (2016). *Advanced Driver-Assistance Systems: Challenges and Opportunities Ahead.* McKinsey & Company.

2 Kamal, R. (2009). *Embedded Systems: Architecture, Programming, and Design*, 2e. McGraw-Hill.

3 Ünsalan, C. et al. (2022). *Embedded System Design with ARM Cortex-M Microcontrollers: Applications with C.* Springer.

4 Amos, B. (2020). *Hands-On RTOS with Microcontrollers: Building Real-Time Embedded Systems Using FreeRTOS, STM32 MCUs, and SEGGER Debug Tools.* Packt Publishing.

5 Gomaa, H. (2016). *Real Time Software Design for Embedded Systems.* Cambridge University Press.

6 Kopetz, H. (2011). *Real-Time Systems: Design Principles for Distributed Embedded Applications.* Springer.

7 Buttazzo, G.C. (2011). *Hard Real-Time Computing Systems: Predictable Scheduling Algorithms and Applications.* Springer.

8 Tanenbaum, A.S. and Bos, H. (2014). *Modern Operating Systems.* Pearson.

9 Smith, J. and Johnson, A. (2021). Hierarchical architecture for embedded systems. *Journal of Embedded Systems* 123–135.

10 Razzaque, M.A. et al. (2016). "Middleware for internet of things: a survey," *IEEE Internet of Things Journal* 70–95.

11 Datta, S.K. et al. (2016). "Vehicles as connected resources: opportunities and challenges for the future," *IEEE Vehicular Technology Conference* 1–5.

12 Friha, O. et al. (2021). Internet of things for the future of smart agriculture: a comprehensive survey of emerging technologies. *IEEE/CAA Journal of Automatica Sinica* 8 (4): 718–752.

13 Fan, X. (2015). *Real-Time Embedded Systems: Design Principles and Engineering Practices*, 1e.

14 Mani, D. (2016). "Implementing finite state machines in embedded systems," *Embedded.*

15 "ARM Cortex-M Processor Documentation," 2024, https://developer.arm.com/ip-products/processors/cortex-m

16 "Nordic nRF5340 Dual-core Bluetooth 5.4 SoC," Product Homepage, https://www.nordicsemi.com/Products/nRF5340

17 Texas Instruments (2024). *MSP432E401Y SimpleLink™ Ethernet Microcontroller Datasheet.* Texas Instruments.

8

Embedded Software Optimization

8.1 Introduction

As the second part of embedded software development, this chapter addresses several key aspects of the field, focusing on optimization, programming languages, development tools, debugging tools, and testing strategies. The optimization of embedded software is crucial for executing tasks in real time with limited Central Processing Unit (CPU) resources. Factors influencing runtime performance, such as memory management, code efficiency, power usage, hardware utilization, and concurrency, are discussed in detail. The optimization process begins with benchmarking to identify system bottlenecks and employs profiling tools to monitor resource consumption, ensuring that all features operate efficiently.

The choice of programming language in embedded software development is influenced by application requirements, including performance, resource constraints, and development speed. C and C++ are frequently preferred for performance-critical tasks due to their efficiency and low-level access capabilities, while Python and MicroPython are recognized for their ease of use and rapid development features. Assembly language is also explored for its ability to provide precise hardware control, making it suitable for tasks that demand high performance.

Development tools are essential for creating, compiling, and testing embedded software. Integrated Development Environments (IDEs), such as Keil MDK and IAR Embedded Workbench, are highlighted for their comprehensive features, including code editors, compilers, and debugging tools. These environments streamline the development process and enhance productivity, making them indispensable for embedded system developers. Debugging tools, both hardware and software, are employed to diagnose software issues effectively. These tools enable the tracing of program flow and verification of functionality while allowing for the monitoring of global variables and memory at runtime. Commonly used debugging tools, such as Joint Test Action Group (JTAG) and Serial Wire Debug

Introduction to Wireless System Design: From Circuits to Web-based Applications, First Edition.
Henry Lau, Ludy Liu, and Keith Chan.

(SWD) debuggers, facilitate real-time debugging by connecting to microcontroller interfaces. In-circuit emulators (ICEs) are also discussed for their deep debugging capabilities, providing access to microcontroller internals.

Finally, embedded software testing is integral to verifying functionality and ensuring the success of system design. Testing strategies encompass unit tests that evaluate individual software modules and system tests that assess overall functionality. Thorough testing of all defined use cases is vital for maintaining system integrity. Errors identified during testing lead to necessary software revisions, enhancing robustness, and reliability. This iterative process contributes to a more resilient embedded system, ensuring that software and hardware components work together effectively and that potential errors are proactively addressed to prevent critical failures.

Through this comprehensive overview, the chapter aims to equip developers with the knowledge and tools necessary for successful embedded software development, ultimately enhancing system performance, and reliability.

8.2 Software Optimization

An application running on an embedded system is designed to execute specific tasks in real time using limited CPU resources. Various factors affect the runtime performance of an embedded system, which are discussed in this chapter. This section covers the process of software optimization (as depicted in Figure 8.1), including techniques for optimizing memory management, code efficiency, power usage, hardware utilization, concurrency, and the importance of benchmarking and testing. The optimization process typically begins with benchmarking to identify system bottlenecks. This involves measuring system performance to ensure that all features, communications, and functionalities operate as expected. For instance, it is crucial that an embedded system does not utilize 100% of its CPU bandwidth under normal operating conditions. Specific requirements for CPU resource usage depend on the overall system design and the applications involved. It is vital to estimate CPU resource needs during the system design or even during the microcontroller (MCU) selection phase.

If a system cannot complete tasks within required time frames due to inadequate CPU resources, it may lead to unexpected outcomes, including system failures or erratic behavior. Many chip providers offer profiling tools within their development environments to monitor CPU resource usage during runtime. Engineers must understand how much CPU resource each software component consumes, allowing for accurate estimation of the total CPU resource usage of the entire system. This includes breaking down CPU resource usage on a use-case basis, facilitating precise estimation of worst-case scenarios in terms of CPU bandwidth (measured in million instructions per second [MIPS] or DMIPS [Dhrystone MIPS]) and memory usage.

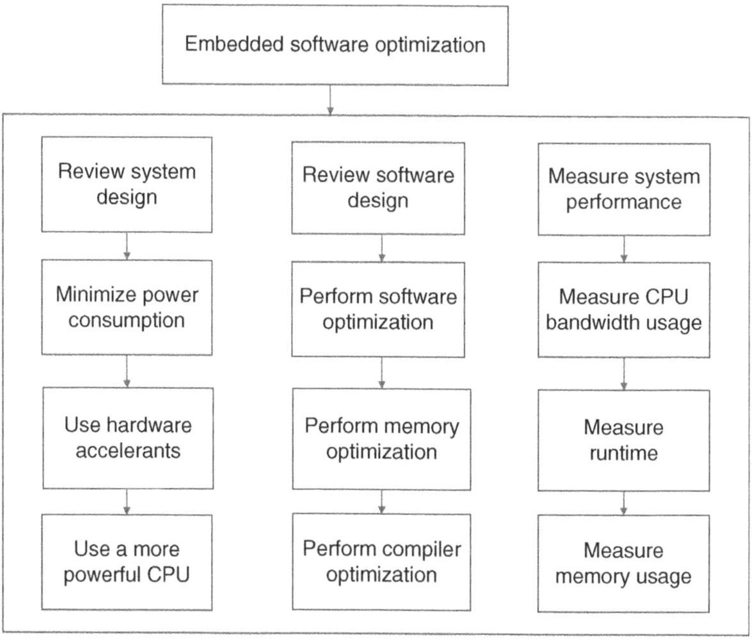

Figure 8.1 Embedded software optimization process

To enhance runtime performance, both the system and software design should be reviewed by engineers. This involves an evaluation of the overall program flow, communication schemes among CPU cores and subsystems, and the data structures used for storing time-critical data. These elements can be optimized to significantly reduce memory access latency and improve efficiency. For example, appropriate data structures can be selected to minimize memory usage and enhance data locality, leading to faster access times. Once the bottleneck is identified through benchmarking, the code should be optimized by engineers. This includes the refinement of algorithms for efficiency, the employment of inline functions, and the unrolling of loops where appropriate to reduce execution time. Additionally, memory management techniques, such as static allocation and memory pools, can be employed to minimize fragmentation and improve access speeds.

Power optimization is another critical aspect. Implementing low-power modes and dynamic frequency scaling can significantly reduce energy consumption. These strategies help ensure that the system can operate efficiently without compromising performance. Utilizing hardware features effectively is also essential. Techniques like Direct Memory Access (DMA) can offload data transfer tasks from the CPU, allowing for more efficient processing. Leveraging hardware accelerators can further enhance performance, particularly for computationally intensive tasks.

Concurrency plays a vital role in optimizing embedded systems. Implementing multithreading allows the system to handle multiple operations simultaneously, enhancing responsiveness. Using a Real-Time Operating System (RTOS) can improve task scheduling and resource management, ensuring that critical tasks meet their deadlines.

The benchmarking and optimization process is iterative, continuing until the desired performance goals are achieved. Testing and validation are crucial throughout this process. Automated testing frameworks can help identify performance regressions, allowing for timely adjustments and refinements.

8.2.1 Benchmarking of an Embedded System

Benchmarking is an effective way to measure the required runtime and CPU bandwidth consumption of application software running on a target system. It involves assessing the runtime of a specific task, individual feature, system component, or the entire system. This process is used to identify bottlenecks within the system and can also facilitate comparisons between the performance of multiple algorithms. Additionally, benchmarking is important for investigating the performance of a MCU during the MCU selection phase. The goal of benchmarking is to ensure that the chosen MCU has sufficient processing power to handle all the specific tasks required for the system under design.

Systematically stressing the system is an effective method for identifying bottlenecks and determining the worst-case runtime performance. This involves identifying a use case that causes the system to consume the maximum CPU resources while performing the specified tasks. Once bottlenecks are identified, optimization can be performed on specific software components or the entire system according to the timing needs.

Benchmarking tools play a crucial role in optimizing embedded systems, as they help effectively identify performance bottlenecks and resource usage. One of the most widely used profiling tools is gprof [1], a GNU (GNU's Not Unix) profiler that analyzes the time consumption of applications. It provides a call graph, allowing developers to see which functions consume the most CPU time. Another versatile option is Valgrind [2], which includes several profiling capabilities: Memcheck for memory profiling, Cachegrind for cache usage analysis, and Callgrind for performance analysis. Valgrind is particularly useful for detecting memory leaks and optimizing memory usage, making it invaluable for embedded software development. For real-time performance analysis, tools like Tracealyzer [3] are excellent choices. This tool visualizes and analyzes the timing behavior of real-time systems, providing insights into task execution times, interruptions, and resource contention. Additionally, FreeRTOS+Trace [4] is specifically designed for applications running on FreeRTOS, helping developers visualize task execution and identify scheduling bottlenecks.

Static analysis tools also contribute significantly to performance optimization. PC-lint [5] serves as a static code analysis tool that detects potential performance issues and coding errors before runtime. Another robust option is Coverity [6], which offers static analysis to uncover bugs and vulnerabilities that can impact performance. When working with embedded systems, using development environment-specific tools can be advantageous. For instance, Keil MDK (Microcontroller Development Kit) [7] is an integrated environment tailored for ARM microcontrollers, featuring built-in analysis tools for performance monitoring and code coverage. Similarly, IAR Embedded Workbench [8] provides profiling tools to optimize code size and execution speed for various microcontroller architectures.

For hardware-based performance monitoring, the ARM Cortex-M Performance Monitoring Unit (PMU) is invaluable if you are using ARM Cortex-M processors. The PMU offers cycle counts, instruction counts, and other essential metrics necessary for optimization. In environments based on Intel architecture, the Intel VTune Profiler [9] offers detailed performance profiling and analysis capabilities. Network performance testing is another critical area for embedded systems that rely on communication protocols. Wireshark [10], an open-source packet analyzer, helps analyze network traffic and diagnose communication issues, while iperf measures the maximum TCP (transmission control protocol) and UDP (user datagram protocol) bandwidth performance, making it useful for optimizing network-related performance.

Additionally, many IDEs like Eclipse [11], MPLAB X [12], and Visual Studio [13] come with integrated profiling and debugging tools that are beneficial for benchmarking embedded applications.

8.2.2 Optimizing Memory Management

Key memory technologies used in today's advanced systems often include Static Random Access Memory (SRAM), Double Data Rate (DDR), and Synchronous Dynamic Random Access Memory (SDRAM). Memory technologies that provide high bandwidth and throughput are required by advanced embedded applications to handle intensive data processing demands. Challenges are often faced by engineers in minimizing memory usage and latency to ensure that sufficient CPU resources are available for performing dedicated tasks in real time. Methods for minimizing memory access latency are thus considered crucial for optimizing system performance.

8.2.2.1 Minimizing Memory Access Latency

Reducing output data size is an effective method for minimizing memory access latency. One common technique for reducing input data size is data compression. A notable example of this concept in embedded wireless systems is found in facial recognition and synthesis. Instead of transmitting a complete image of a face, a general model is utilized that describes human faces using anatomical parameters such as nose shape, eye separation, skin color, and cheekbone angle.

By sending these parameters over wireless networks, the face can be recreated within the model, significantly reducing the data required for transmission. This approach is especially advantageous in embedded systems used in video conferencing, smart cameras, and augmented reality, where efficient data transfer and low bandwidth usage are crucial.

8.2.2.2 Speeding Up Memory Access for Critical Tasks

To improve the runtime performance of the application software, optimizing algorithms and/or data structures may be required for some software modules. For example, critical tasks with strict timing requirements should have access to faster memory, such as SRAM, to minimize latency. A well-defined memory scheme benefits runtime performance. Different memory technologies and their characteristics and performance will be discussed.

8.2.2.3 SDRAM, SRAM, or DDR

Choosing between SDRAM, SRAM, and DDR for embedded systems is influenced by a variety of factors, including cost, power consumption, performance requirements, and specific application needs. SDRAM is known for offering good performance for applications that require moderate data bandwidth. It is generally regarded as more cost-effective than SRAM, making it suitable for general-purpose embedded systems. However, SDRAM requires periodic refreshing, which can introduce latency, and higher power consumption is typically associated with SDRAM compared to SRAM.

SRAM is characterized by faster access times than SDRAM and DDR, making it an excellent choice for applications where speed is critical. No refreshing is required, which helps reduce latency and enhance reliability in time-sensitive applications. However, higher costs are generally associated with SRAM, and lower density means less storage capacity is provided per chip. This increased cost can be a significant consideration for projects with tight budgets or large-scale deployments. DDR memory is recognized for providing higher data transfer rates compared to SDRAM, making it suitable for high-performance applications. A good balance between speed and efficiency is struck, particularly in later generations like DDR3 and DDR4. However, a more complex interface and control requirements are typically involved with DDR, and higher power consumption may be associated with it compared to SDRAM for similar capacities. The cost of DDR can vary significantly based on generation and technology, which should be taken into account when considering overall budget constraints.

When these options are considered for embedded systems, the specific performance needs of the application should be assessed. If high speed and low latency are demanded by the system, SRAM might be viewed as the best choice. For moderate performance with a focus on cost, SDRAM is often seen as suitable. In contrast, DDR is regarded as ideal for high-performance applications requiring greater bandwidth. Power consumption is also recognized as a critical factor, especially for battery-operated or energy-sensitive devices. In such cases, SRAM may be preferred due to lower power usage during standby. Cost considerations play a significant role as well; while SDRAM is often viewed as the most budget-friendly option, the overall cost-effectiveness will depend on the specific requirements and scale of the project.

8.2.2.4 Bridging the Performance Gap Between SDRAM and SRAM

In modern processor architectures, such as ARM, the memory hierarchy typically consists of multiple levels of cache [14, 15]. This structure bridges the performance gap between the processor and main memory by temporarily storing small sections of code in cache memory closer to the processor. Caches operate on two key principles of locality in memory access. Spatial locality [15] means that when one memory location is accessed, nearby locations are likely to be accessed soon after, which is often observed in data structure usage and during sequential instruction execution. Temporal locality indicates that if a specific memory area is accessed, it is likely to be accessed again in a short time frame, as occurs during the repeated execution of loops in code. To minimize control information storage, spatial locality groups multiple memory locations under a single tag [15], creating a cache line. Loading data into a cache improves subsequent access times, leading to better overall performance.

Most modern processors provide multiple levels of caches: L1, L2, and L3 caches. L1 cache is the closest and fastest cache to the processor core, typically implemented using SRAM technology. It is divided into a program cache for storing software programs and a data cache for storing data. L1 cache has the lowest latency compared to L2 and L3 caches and has a very small memory footprint. L2 cache is the second level of cache, also typically implemented using SRAM technology. L2 cache is larger than L1 cache in terms of memory footprint and has higher latency. However, it still provides faster access than the main memory, such as SDRAM. L3 cache is the third level of cache. It can be implemented using either SRAM or a combination of SRAM and SDRAM. L3 cache has a larger memory footprint than L2 cache and higher latency. However, it can still significantly improve overall system performance.

Consider now the impact of the L1 cache on the overall performance of a computer system [15]. Let h be the hit rate, M the miss penalty (the time to access information in the main memory), and C the time to access information in the L1 cache. The average access time experienced by the processor is

$$t_{\text{ave}} = (C \times h) + M \times (1 - h) \tag{8.1}$$

This formula illustrates how the L1 cache hit rate and the respective access times influence overall performance. Including both L1 and L2 caches further reduces the impact of main memory speed on the performance of a computer.

Now consider the impact of the L2 cache on the overall performance of a computer system. Let h_1 and h_2 be the hit rates, and M_1 and M_2 be the miss penalties (the time to access information in the main memory), with C_1 and C_2 being the times to access information in the L1 and L2 caches, respectively.

The average access time experienced by the processor in a system with two levels of caches is

$$t_{\text{ave2}} = (C_1 \times h_1) + (1 - h_1) \times (h_2 \times C_2) + (h_1 \times M_2) \times (1 - h_2) \times M_1 \tag{8.2}$$

This formula illustrates how the L1 and L2 cache hit rates and their respective access times influence overall performance.

8.2.2.5 Tightly-Coupled Memory

For latency-sensitive systems, consider using Tightly-Coupled Memory (TCM) [16], which is often found in ARM-based architectures. TCM provides very low access latency, typically within a single processor cycle. Due to the smaller memory footprint and higher cost, it is often used to support the most time-critical code and data, such as interrupt handlers and DMA transfers. Figure 8.2 shows an example of the mapping of system memory to TCM in an ARM Cortex-M7-based architecture. The TCM port features Instruction TCM (ITCM) and Data TCM interfaces.

In a real-time embedded system, time-critical sensor data, such as audio or camera data, must be transferred to memory buffers as soon as it is available to the peripheral. The code design for supporting data transfer must be optimized to ensure that time-critical data is accessed with minimal latency. Consider storing the sensor data buffer in TCM to ensure fast, low-latency access to the data. The sensor data should be transferred from the peripheral to memory using DMA without involving the CPU. To ensure that the data is moved to the internal memory buffer as soon as it becomes available, the DMA transfer should be configured to be triggered by an interrupt upon the arrival of sensor data.

8.2.3 Optimizing Data Structure

Regardless of whether specific data is stored in fast or slow memory, optimizing data structures is a critical aspect of embedded software development. Given the constraints of embedded systems, such as limited memory, processing power, and

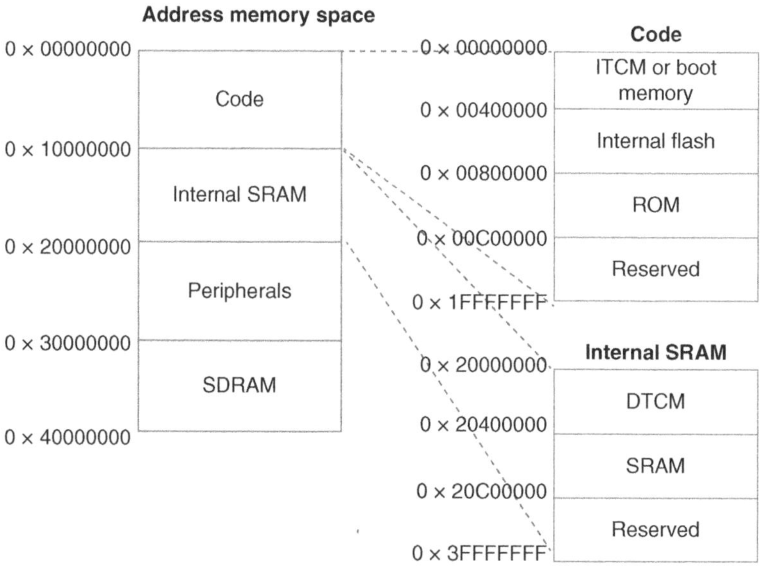

Figure 8.2 Mapping of system memory to TCM in an Arm Cortex-M7-based architecture

energy resources, selecting and designing efficient data structures can significantly impact performance, memory usage, and overall system responsiveness. The choice of data structure should be tailored to the specific use case, with common options including arrays, linked lists, hash tables, and trees. For instance, arrays are ideal for fixed-size collections that require fast random access, while linked lists are useful for dynamic memory allocation, allowing efficient insertions and deletions. However, linked lists introduce overhead due to pointer storage and may lead to cache inefficiencies. Hash tables can provide average-case constant-time complexity for lookups and insertions, making them effective for scenarios requiring quick access to elements, although they may require more memory and can suffer from collisions. By analyzing the specific access patterns and operational requirements, developers can select the most suitable data structure.

The memory layout and alignment of data structures also play a crucial role in performance. Using contiguous memory allocation, such as arrays or structures that store elements sequentially, can improve cache locality and lead to faster access times. This is particularly important in embedded systems where cache performance significantly affects overall speed. Additionally, ensuring that data structures are properly aligned according to the architecture's requirements can prevent performance penalties associated with misalignment. Developers can utilize compiler directives or attributes to enforce proper alignment, thereby enhancing efficiency.

Minimizing the memory footprint of data structures is essential in resource-constrained environments. One effective strategy is to use fixed-size data types, such as `int8_t` or `int16_t`, instead of generic types like `int`, which can vary in size depending on the architecture. Implementing bit fields within structures allows multiple Boolean or small integer values to be packed into a single byte, thus reducing overall memory usage. Moreover, favoring static allocation over dynamic allocation can help avoid fragmentation and unpredictable memory usage patterns, leading to more efficient memory management.

Access patterns significantly influence the efficiency of data structures, making it important to optimize them accordingly. Organizing data structures to maximize spatial and temporal locality can enhance performance; for example, grouping-related data together or using structures of arrays (SoA) instead of arrays of structures (AoS) can improve cache utilization. Sequential access patterns should also be encouraged, as they can lead to better cache performance and fewer cache misses. The choice of algorithms used to manipulate data structures is equally crucial. Lightweight iterators or pointers can be employed to efficiently traverse data structures, while avoiding unnecessary copying of data. Analyzing the time complexity of operations on data structures is essential to ensure the chosen algorithms meet the performance requirements of the system.

Regular profiling and testing of data structure performance can help identify inefficiencies, making it vital to incorporate these practices into the development process. Profiling tools can be used to measure the performance of different data structures under real-world conditions, helping to pinpoint bottlenecks and inform decisions about potential optimizations. Implementing benchmarks for critical data structure operations ensures they meet the required performance criteria.

In multithreaded embedded systems, data structure design must also accommodate concurrency. Utilizing lock-free or wait-free data structures can minimize contention between threads, which is essential for maintaining performance. Additionally, considering thread-local storage for data structures that are frequently accessed by individual threads can further reduce contention and improve access speed.

8.2.4 Minimizing Power Consumption

CPU power consumption is directly related to the amount of work the processor is doing. Minimizing CPU and memory usage is an effective method for reducing power consumption. Power consumption is influenced by memory usage, as accessing and maintaining data in memory requires energy. Reducing power consumption is achieved by minimizing memory usage for data and code, and one way to lower power consumption is to reduce unnecessary CPU usage. In addition to using state-of-the-art memory technology and optimizing data structures, designing efficient software can help. Another approach is to offload some system tasks to coprocessors or hardware accelerators. A common method to reduce power consumption in embedded systems is to put the CPU into low-power or sleep mode.

8.2.4.1 Low-Power Mode

Sleep mode, also known as low-power mode or idle mode, is a power-saving state in which the processor and other system components are put into a low-power state and temporarily forced into sleep or idle mode. This helps to reduce power consumption during sleep/idle periods by performing only the mandatory safety operations such as monitoring the temperature sensors to prevent overheating of hardware. The system is designed to wake up by external events or interrupts. Sleep mode is often implemented in embedded systems to optimize power consumption, especially in battery-powered devices where extending battery life is crucial. As discussed in the previous chapter, designing a tracking watch presents unique challenges compared to other consumer products, primarily due to its compact size and limited battery capacity. To address these challenges, smart tracking watches can be engineered to conserve energy through efficient wake-up mechanisms and sleep modes.

Motion sensors and other detection devices are strategically integrated into the watch to identify triggering events, such as user movement or changes in ambient air quality. When these sensors detect an event, they generate an interrupt signal that is immediately processed by the watch's microcontroller or embedded processor. This interrupt-driven approach enables a swift transition from low-power sleep mode to an active state. Instead of remaining in a constant active state, the watch can enter sleep mode when no activity or alerts are detected. It will only activate and adjust its display or tracking features in response to sensor-triggered interrupts, ensuring that power is consumed only when necessary.

8.2.4.2 Watchdog Timer

Watchdog timers [17] play a crucial role in optimizing energy consumption and CPU utilization within embedded systems. Power consumption and CPU bandwidth can be reduced by using a watchdog timer to monitor the system's responsiveness. A watchdog timer can be used to place a system into sleep mode and wake it up when needed. By continuously monitoring the system's operation, the watchdog timer allows the processor and other components to enter low-power modes when there is no immediate need for active processing. When the system is idle or performing noncritical tasks, the watchdog timer tracks the system's responsiveness, resetting its internal counter as long as the software is functioning properly. However, if the software fails to reset the watchdog timer, indicating a potential system failure or lack of response, the watchdog timer will trigger an

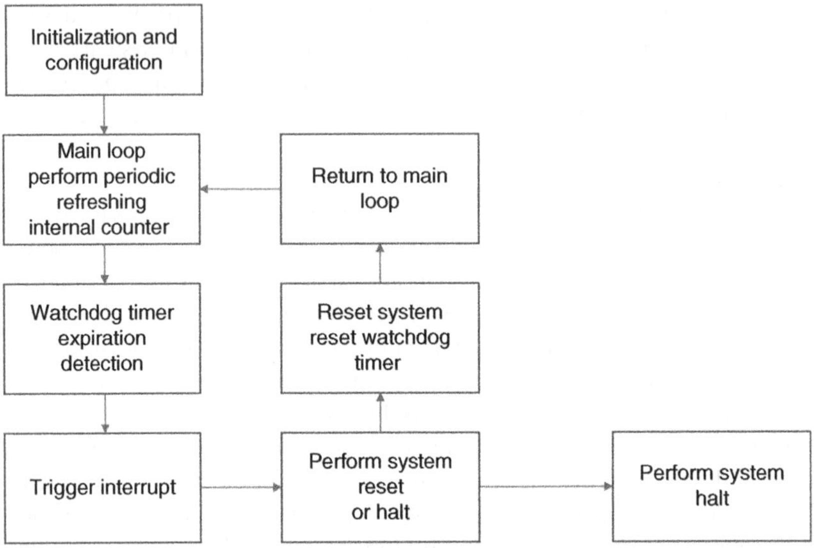

Figure 8.3 Program flow of a watchdog timer to monitor system responsiveness

interrupt, waking the processor from its low-power state. This interrupt-driven approach ensures that the system can quickly transition back to an active state to address any issues without requiring the processor to remain in a high-power mode constantly. Figure 8.3 shows the program flow of using a watchdog timer to monitor the responsiveness of the main loop.

8.2.5 Leveraging Hardware Accelerator

A hardware accelerator is a specialized hardware component designed to accelerate computationally intensive tasks or operations, significantly enhancing the performance of embedded systems. These accelerators offload specific processing tasks from the main CPU, allowing it to focus on general-purpose computing while the accelerator handles specialized functions more efficiently. Some prominent examples of hardware accelerators include filter accelerators, Fast Fourier Transform (FFT) accelerators, network processors, floating-point accelerators, and fixed-point accelerators.

One of the primary benefits of using hardware accelerators is the increased processing speed they offer. Optimized for specific algorithms, these components can execute tasks much faster than general-purpose CPUs. For example, a filter accelerator can process real-time audio or video data more swiftly, improving the responsiveness of applications that rely on these signals. This speed advantage translates into reduced latency, which is critical in applications where timely responses are essential, such as robotics or real-time control systems.

In addition to enhancing speed, hardware accelerators can lower power consumption. They are often designed to perform tasks using less power compared to a CPU running at full capacity. This efficiency not only extends the battery life of portable devices but also reduces heat generation, which is crucial in compact embedded systems. Furthermore, many hardware accelerators can execute multiple operations simultaneously, allowing for enhanced parallel processing. For instance, network processors can handle multiple data streams concurrently, which is vital for applications like telecommunications and data processing.

The use of hardware accelerators is also optimized for resource utilization within an embedded system. Specific tasks are delegated to these dedicated components, allowing more complex algorithms to be executed without overloading the primary processor. This optimization results in a more efficient system overall. Additionally, with specific tasks being managed by hardware accelerators, software can be designed with a clearer focus on high-level functionality rather than low-level optimization. This separation simplifies development and maintenance, enabling faster iterations and updates. Moreover, improved system scalability is contributed to by the integration of hardware accelerators. As application demands grow, existing systems can have accelerators added to enhance performance without requiring a complete redesign. This modular approach allows for scalable solutions that can adapt to changing requirements. Many hardware accelerators are also tailored for specialized applications, such as image processing or encryption, enabling algorithms to be implemented more efficiently than by general-purpose CPUs.

8.2.5.1 Filter Accelerators

Filter accelerators are frequently integrated into embedded systems to alleviate the computational burden of intensive signal processing tasks, such as audio and video processing, from the main CPU. By taking on these specialized operations, filter accelerators significantly improve the overall performance and efficiency of the system. They are especially beneficial in applications that demand real-time processing, where any delays could negatively impact user experience or the functionality of the system.

These accelerators are typically implemented in system-on-chip (SoC) designs, working alongside the CPU to deliver efficient and low-latency processing capabilities. This collaboration enables the CPU to concentrate on higher-level tasks while the filter accelerator specializes in executing specific algorithms related to signal processing. For example, in mobile devices, filter accelerators handle essential functions like noise reduction, echo cancellation, and image stabilization, which are critical for smooth audio and video playback. A prominent example of a filter accelerator is the Qualcomm Hexagon DSP [18], which is integrated into Snapdragon SoCs. This digital signal processor is specifically engineered to manage a variety of multimedia tasks, facilitating high-quality audio and video processing while optimizing power consumption. By leveraging the Hexagon DSP, devices can achieve enhanced performance in applications such as video conferencing, gaming, and augmented reality, where real-time processing is essential.

8.2.5.2 Network Processors

Network processors are specialized components designed for high-performance network processing tasks, crucial for the efficient operation of modern embedded systems. These processors are integrated into various devices to manage high-speed network operations, including packet processing, traffic management, and protocol handling. By offloading these demanding network-related tasks from the main CPU, network processors significantly enhance overall system performance and efficiency. These processors are particularly valuable in environments where rapid data transmission and processing are essential. They are commonly found in network equipment such as routers, switches, and firewalls, where they facilitate the seamless flow of information across networks. In addition, network processors are integral to industrial automation systems, where they help manage communication between devices and sensors, ensuring timely data exchange and control. They are also increasingly utilized in edge computing devices, which require real-time data processing close to the source of data generation, thereby minimizing latency and bandwidth usage.

An exemplary model of a network processor is the Intel Ethernet Controller X710 [19], designed for 10GbE (10 Gigabit Ethernet) networks. This controller is engineered to support high-speed data transfers and advanced features like virtualization, which allows multiple virtual machines to share the same network interface without compromising performance. By utilizing the X710, organizations can achieve enhanced network throughput and reduced latency, making it ideal for data-intensive applications, such as cloud computing, video streaming, and large-scale data analysis.

8.2.5.3 Floating-Point Units

Floating-point units (FPUs) [20] are specialized hardware implemented on SoCs to accelerate floating-point arithmetic operations. FPUs are integrated into embedded systems, including SoC designs, to enhance the performance of floating-point intensive workloads, such as those found in scientific computing, graphics processing, and machine learning applications. On processors without an FPU, these operations are handled by software via the C compiler's standard library, resulting in significantly lower performance. Floating-point calculations can be resource-intensive, as each operation between two numbers involves several steps. Initially, the numbers must be aligned by adjusting them to have the same exponent. Next, the actual mathematical operation is performed, followed by rounding the result to fit within the specified precision. Finally, the result is encoded for output or storage. While the software-based approach makes these processes invisible to the programmer, the performance hit can be substantial.

In contrast, processors equipped with an FPU can handle these operations directly in hardware, typically executing most instructions in a single cycle. In this case, the C compiler generates native FPU instructions rather than relying on its own floating-point library. When implementing mathematical algorithms on a microprocessor with an FPU, programmers benefit from both performance and ease of development. The FPU enhances reliability by allowing the use of high-level tools to generate code optimized for the hardware, eliminating the need to compromise between speed and development efficiency.

8.2.5.4 Fast Fourier Transform Accelerators

FFT accelerators [21, 22] are specialized for accelerating the computation of FFTs. FFT accelerators are integrated into embedded systems that are used in signal processing applications, such as audio and voice processing, communications, radar signal processing, and image/video processing. It is important to determine whether FFT accelerators are available on a specific processor during the system design phase or the processor selection phase. Many modern SoC systems have hardware accelerators that can be utilized to support specific complex mathematical operations, such as FFT and Inverse Fast Fourier Transform calculations, as well as other floating-point operations like sine, cosine, and floating-point division efficiently.

The use of these hardware accelerators can greatly reduce software complexity, memory usage, and memory access latency. A significant portion of the processing burden can be offloaded from the main CPU by these specialized hardware components. Attention must be paid when assigning a core to run specific functions or features. The capabilities and features offered by a specific core in a multicore SoC system must be fully understood by engineers and architects.

8.2.5.5 Fixed-Point Accelerators

Fixed-point accelerators are specialized hardware designed to accelerate fixed-point arithmetic operations. They are commonly integrated into embedded systems, particularly in resource-constrained environments like Internet of Things (IoT) devices, industrial automation, and control systems. These accelerators leverage specialized hardware blocks and Single Instruction, Multiple Data (SIMD) instructions to efficiently execute fixed-point arithmetic operations, offloading a significant portion of the processing load from the main CPU and improving the overall performance and energy efficiency of the embedded system [23]. For example, ARM Cortex-M processors with SIMD extensions can reduce CPU bandwidth usage for fixed-point operations in embedded systems.

8.2.5.6 Multicore Accelerators

Multicore accelerators are designed to leverage multiple processing cores for the parallel execution of tasks. These accelerators are integrated into embedded systems to provide increased processing power and parallelism for computationally intensive tasks. By offloading a significant portion of the processing load from the main CPU, multicore accelerators can improve the real-time capabilities and overall performance of embedded systems [24]. Other examples of multicore accelerators widely used in various embedded and IoT applications are discussed in the next few sections. These multicore accelerators are often integrated alongside the main microcontroller or microprocessor in IoT devices, forming a heterogeneous computing architecture. This allows the system to offload computationally intensive tasks to the specialized cores, while the main processor handles overall system coordination and higher-level application logic.

8.3 Programming Languages

The choice of programming language in embedded software development often depends on the specific requirements of the application, including performance, resource constraints, and development speed. C and C++ remain the dominant choices for performance-critical tasks, while languages like Python and MicroPython are valued for their ease of use and rapid development capabilities.

Assembly language represents a low-level programming approach that provides a symbolic representation of machine code specific to a particular CPU architecture. This language offers precise control over hardware, allowing developers to optimize for speed and memory usage. Although writing in assembly can lead to extremely efficient programs, it tends to be more complex and harder to maintain. Assembly is often used in performance-critical applications, bootloaders, or in scenarios requiring direct hardware manipulation, such as embedded systems in aerospace or medical devices.

C is a general-purpose programming language that provides low-level access to memory and hardware, making it highly suitable for embedded systems. One of its key features is efficiency; C code is typically compiled to machine code, allowing for high performance and minimal resource usage. Despite being low-level, C retains a degree of portability, enabling developers to use it across different platforms with minimal changes. The language also benefits from extensive libraries that facilitate various tasks, including hardware interfacing and real-time operations. C is commonly employed in microcontrollers, real-time systems, and applications requiring direct hardware manipulation, such as device drivers and operating systems. As an extension of C, C++ introduces object-oriented programming (OOP) features that allow for more complex system design and better code organization. Its key characteristics include support for classes and objects, which enable encapsulation, inheritance, and polymorphism. Additionally, C++ includes the Standard Template Library (STL), offering a rich set of algorithms and data structures. While it provides manual memory management through constructors and destructors, C++ is particularly suitable for larger embedded systems and applications with complex architectures where performance is critical, such as robotics and automotive systems.

Python is a high-level programming language known for its readability and simplicity, making it popular for a wide range of applications. Its ease of use, characterized by a simple syntax and dynamic typing, allows for rapid development. Python boasts extensive standard libraries and third-party modules, such as NumPy and SciPy, which facilitate various tasks. Additionally, it is cross-platform, enabling it to run on multiple environments, making it versatile for different embedded systems. Python is commonly used in higher-level applications, prototyping, and scripting tasks, particularly in IoT devices for data processing and communication. As a lean implementation of Python, MicroPython is designed specifically to run on microcontrollers and in constrained environments. It is characterized by its small footprint, which allows it to occupy minimal memory, making it suitable for microcontrollers. MicroPython retains much of Python's syntax and functionality, allowing developers to leverage their existing Python knowledge. It also supports a Read-Eval-Print Loop for interactive programming, which is particularly useful for testing and debugging. This language is ideal for IoT applications, hobbyist projects, and educational purposes, and it is commonly used on platforms like the ESP8266, ESP32, and Raspberry Pi Pico.

8.4 Development Tools

Development tools are essential for creating, compiling, and testing embedded software. IDEs are commonly used in this field, as they typically encompass a code editor, compiler, and debugger all within a single application. Many chip manufacturers provide their own IDEs, which often include support for both emulators and debuggers. Emulators allow developers to run their software in a simulated environment, even when the actual target hardware is unavailable. This overview highlights some of the widely used IDEs (some of which have been introduced in the Debugging section) in embedded software development.

Keil MDK, as introduced earlier, is a comprehensive development environment tailored specifically for ARM Cortex-M microcontrollers. It features an intuitive IDE called μVision, which streamlines project management, code editing, and debugging tasks. The high-performance C/C++ compiler included in Keil MDK is optimized for ARM architecture, ensuring efficient code generation. Additionally, the integrated debugger supports real-time debugging, enabling developers to set breakpoints, monitor variables, and execute code step-by-step. Keil MDK also supports various RTOS, making it suitable for complex embedded applications. With its middleware libraries for USB, TCP/IP, and file systems, Keil MDK is ideal for professional developers working in industries such as automotive, medical, and consumer electronics.

IAR Embedded Workbench is another powerful IDE recognized for its high-performance optimizing compiler and versatile debugging tools. The compiler is designed to generate efficient code, which is crucial in resource-constrained environments. The IDE offers comprehensive debugging capabilities, including both source-level and assembly-level features, along with real-time data visualization and performance analysis. IAR Embedded Workbench is compatible with a wide variety of microcontrollers from different manufacturers, enhancing its versatility across projects. Furthermore, it includes tools for static code analysis, helping developers identify potential issues early in the development process. This IDE is particularly favored for safety-critical applications, such as those in aerospace and medical devices, due to its reliability and compliance with industry standards.

Eclipse is an open-source IDE that is highly customizable and widely used across various development environments, including embedded systems. Its plugin architecture allows developers to tailor the IDE to meet their specific needs, with plugins like Eclipse C/C++ Development Tooling commonly utilized for embedded development. Eclipse is cross-platform, functioning on Windows, macOS, and Linux, which makes it accessible for diverse development environments. The IDE supports various build systems and toolchains, including GCC and makefiles, which are essential for embedded projects. Additionally, Eclipse can integrate with version control systems like Git, facilitating team collaboration and project management. This flexibility makes Eclipse particularly appealing to developers who prefer an extensible and open-source solution.

The Arduino IDE [25] is specifically designed for developing applications on Arduino boards and compatible hardware. It features a user-friendly interface that makes it accessible to beginners and hobbyists alike. One of the standout features of the Arduino IDE is its extensive library support, which simplifies tasks such as sensor integration and communication protocols. The IDE allows for straightforward code uploading to Arduino boards with a single click, streamlining the development process. With a large community of users, the Arduino ecosystem offers ample tutorials, forums, and shared projects, aiding in learning and problem-solving.

The Arduino IDE is excellent for rapid prototyping, educational purposes, and hobbyist projects, enabling users to quickly develop and test their ideas.

Mbed Studio [26] is an IDE specifically tailored for developing IoT applications on ARM Cortex-M microcontrollers. It provides a user-friendly interface that combines code editing, project management, and debugging capabilities. Mbed Studio integrates seamlessly with Mbed OS, a free, open-source operating system designed for the IoT, making it easy to develop connected applications. The IDE includes a library manager that simplifies the addition and management of Mbed libraries and dependencies. Furthermore, Mbed Studio supports cloud-based development and deployment, which is particularly beneficial for IoT project management. This IDE is ideal for developers focused on IoT applications, offering tools and support that cater to the needs of modern embedded systems.

8.5 Debugging Tools

Debugging tools, whether hardware or software, are essential for in-depth analysis when diagnosing software issues. These tools are commonly employed to trace program flow and verify that the software operates as intended during the initial debugging phases. They enable developers to monitor global variables, memory contents, and register values at runtime. However, it is important to recognize that halting the code can disrupt its operation. While debuggers excel in analyzing low-level device operations or code execution within specific time frames, they can struggle to debug entire systems, particularly in real-time environments. For comprehensive system testing, alternative approaches are often more suitable. In the event of a software crash, debugging tools can provide valuable insights, such as capturing the call stack at the moment of failure. By allowing developers to step through code line by line, set breakpoints, and examine memory and register states, these tools help pinpoint the root causes of issues.

Among the most commonly used debugging tools are JTAG [27] and SWD [28] debuggers. These standardized protocols facilitate connections to a microcontroller's debug interface, allowing for real-time debugging capabilities. Developers can set breakpoints, step through code, and inspect or modify memory and registers. Notable examples include J-Link, recognized for its speed and compatibility, and ST-Link [29], which is tailored for STM32 microcontrollers.

ICEs also play a vital role in embedded software development. These hardware tools emulate the target microcontroller and provide deep debugging capabilities that surpass those of traditional debuggers. ICEs grant full access to the microcontroller's internals and can simulate various hardware conditions, which is particularly useful for testing edge cases within complex systems where hardware–software interactions are critical.

Serial debugging offers a straightforward yet effective method for communication, utilizing UART. This approach allows developers to print debug messages, error logs, and variable values to a terminal, making it especially useful for systems with limited debugging capabilities or where JTAG/SWD connections are impractical. Logic analyzers and oscilloscopes are crucial instruments for visualizing and analyzing electronic signals in embedded systems. Logic analyzers capture and display the timing of digital signals, aiding in the diagnosis of protocol issues and timing problems, particularly in communication protocols like SPI, I2C, and UART. Oscilloscopes, in contrast, visualize electrical signals as waveforms, enabling developers to measure voltage levels and signal integrity, which is essential for diagnosing analog signal issues.

IDEs, as discussed earlier, are comprehensive platforms that offer a suite of tools for software development, including integrated debugging capabilities. IDEs like Eclipse can be extended with plugins for embedded systems, while Keil is specifically designed for ARM microcontrollers, combining software development, and debugging features in one package. Static analysis tools provide an additional layer of assurance by examining source code for potential errors without execution. These tools help identify bugs, vulnerabilities, and compliance with coding standards. Examples include PC-lint, which checks C/C++ code for issues and violations, and Coverity, which offers in-depth static analysis.

Software simulators are another important resource for embedded developers. These applications run on host machines and replicate the behavior of embedded systems, enabling analysis of software interactions without the need for physical hardware. Popular examples include QEMU [30], an open-source emulator supporting multiple architectures, and Simulink [31], a graphical programming environment designed for modeling and simulating dynamic systems.

8.6 Embedded Software Testing Strategies

Embedded software testing is crucial not only for verifying software functionality but also for ensuring the success of the overall embedded system design. This testing is based on both functional and nonfunctional specifications, aiming to confirm that the software operates as intended within the system. Embedded software testing encompasses unit tests and system tests, as depicted in Figure 8.4. A unit test evaluates individual software modules, verifying that each module aligns with the design specifications. It is essential that all use cases defined during the system definition phase are thoroughly tested, as this contributes to the integrity of the entire system.

Subsystems can be tested separately to confirm that they function as designed, which is vital for the overall reliability of the system. For simple embedded systems, the most basic test involves stepping through the software line by line using

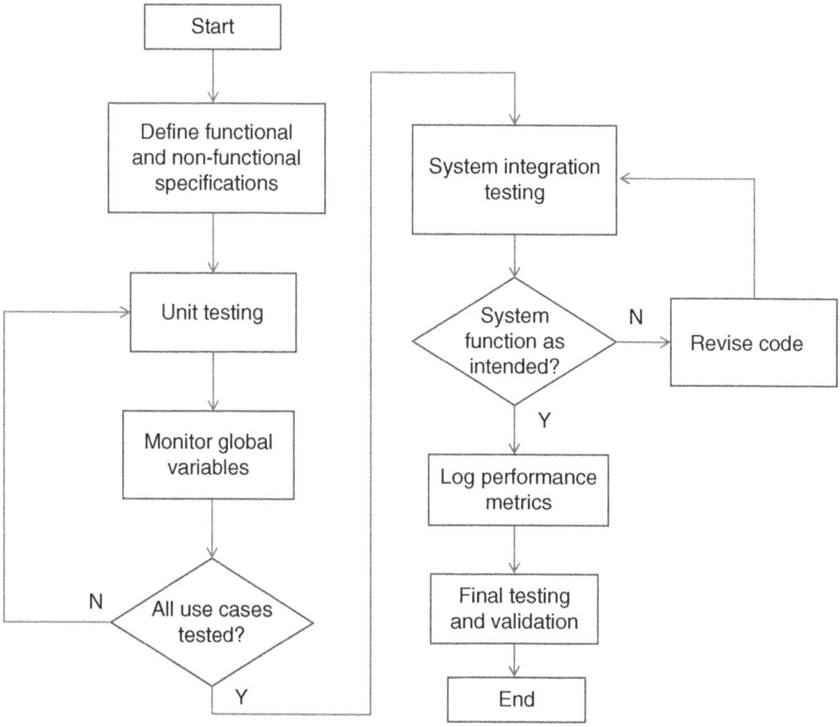

Figure 8.4 Flow diagram of software testing

a debugger. At each step, developers should monitor global variables to ensure the program behaves as expected. In more complex systems, unit tests can be employed to evaluate the smallest software units, with tests crafted based on software requirement specifications. The goal of unit testing is to validate that individual software units perform correctly, thereby supporting the overall system design. Once unit testing is complete, system integration and testing can commence. A system integration test involves assembling the entire system according to specifications and performing tests on the system as a whole. The objective of system testing is to verify that the entire system functions as intended, reinforcing the design's effectiveness. A measurement system may be used to log functional performance, timing performance, and system performance under stress, providing critical insights into system behavior. This measurement system may include a server, oscilloscope, logic analyzer, computer system, network performance analyzers, and custom-designed visualization tools, among others.

Importantly, any errors identified during the testing phases necessitate a revision of the software to implement appropriate error handling and corrective measures. Unhandled errors can lead to severe consequences, such as system hangs or crashes, which compromise the reliability and functionality of the embedded system. This iterative process not only addresses the specific issues discovered but also enhances the robustness of the software, ensuring that it can gracefully handle similar errors in the future. By revising the software based on testing outcomes, developers can improve system reliability and performance, ultimately contributing to a more resilient embedded system.

Embedded software testing is thus integral to the overall design of embedded systems, ensuring that both software and hardware components work harmoniously to meet design specifications and user requirements. The incorporation of error handling through software revisions further strengthens the system, making it more reliable and effective in real-world applications. By addressing potential errors proactively, developers can prevent critical failures that may disrupt system operation.

References

1. "Gprof," Wikipedia, last modified October 7, 2024, https://en.wikipedia.org/wiki/Gprof
2. "Valgrind," last modified October 7, 2024, https://valgrind.org/
3. Percepio, "Tracealyzer," product homepage, https://percepio.com/tracealyzer/
4. FreeRTOS, "TracealyzerTM," product homepage, https://www.freertos.org/Documentation/03-Libraries/02-FreeRTOS-plus/FreeRTOS-Plus/FreeRTOS_Plus_Trace/FreeRTOS_Plus_Trace
5. Vector Informatik GmbH, "Static Code Analysis for C and C++," product homepage, https://pclintplus.com/
6. Synopsys, Inc., "Coverity Scan Static Analysis," product homepage, https://scan.coverity.com/
7. Arm Limited, "Keil MDK, Microcontroller Development Kit," product homepage,https://www.keil.com/
8. IAR Systems AB, "IAR Embedded Workbench," product homepage, https://www.iar.com/
9. Intel Corporation, "Intel VTune Profiler," product homepage, https://www.intel.com/content/www/us/en/developer/tools/oneapi/vtune-profiler.html#gs.frqxru
10. Wireshark Foundation, "Wireshark: Network Protocol Analyzer," product homepage, https://www.wireshark.org/
11. Eclipse Foundation AISBL, "Eclipse IDE for Embedded C/C++ Developers," product homepage, https://www.eclipse.org/downloads/packages/release/2021-03/m1/eclipse-ide-embedded-cc-developers

12 Microchip Technology Inc., "MPLAB® X Integrated Development Environment," product homepage, https://www.microchip.com/en-us/tools-resources/develop/mplab-x-ide

13 Microsoft, "Visual Studio 2022," product homepage, https://visualstudio.microsoft.com/#vs-section

14 Arm Limited, "Arm® v7-M Architecture Reference Manual," https://developer.arm.com/documentation/ddi0403/latest/

15 Hamacher, C., Vranesic, Z., and Zaky, S. (2011). *Computer Organization*, 5e. McGraw-Hill Higher Education.

16 Microchip Technology Inc., "How to Use TCM on Cortext Based MCU with XC32 Compiler," DS90003240, https://ww1.microchip.com/downloads/aemDocuments/documents/MCT32/ProductDocuments/ProductBrief/How_to_Use_TCM_on_Cortex-Based_MCU_with_XC32Compiler_DS90003240A.pdf

17 Texas Instrument, "What Is a Watchdog Timer and Why Is It Important?" Technical Article, https://www.ti.com/lit/ta/ssztah7/ssztah7.pdf?ts=1728573366445&ref_url=https%253A%252F%252Fwww.google.com%252F

18 "Qualcomm Hexagon," Wikipedia, last modified October 7, 2024, https://en.wikipedia.org/wiki/Qualcomm_Hexagon

19 Intel Corporation, "Intel® Ethernet Controller X710," product homepage, https://www.intel.com/content/www/us/en/products/details/ethernet/700-controllers/x710-controllers/downloads.html

20 STMicroElectronics, "Floating point unit demonstration on STM32 microcontrollers," AN4044 Application Note, DocID022737 Revised May 2, 2016, https://www.st.com/resource/en/application_note/dm00047230-floating-point-unit-demonstration-on-stm32-microcontrollers-stmicroelectronics.pdf

21 Brigham, E.O. (1998). *The Fast Fourier Transform and Its Applications.* Prentice-Hall.

22 (2011). *KeyStone Architecture Fast Fourier Transform Coprocessor (FFTC).* SPRUGS2C, Texas Instrument https://www.ti.com/lit/ug/sprugs2c/sprugs2c.pdf?ts=1727156508472&ref_url=htt18]ps%253A%252F%252Fwww.bing.com%252F.

23 Kamal, R. (2008). *Embedded Systems: Architecture, Programming, and Design.* McGraw-Hill Education.

24 Culler, D., Singh, J.P., and Gupta, A. (1999). *Parallel Computer Architecture: A Hardware/Software Approach.* Morgan Kaufmann.

25 Arduino, "Arduino IDE," product homepage, https://www.arduino.cc/en/software

26 Arm Limited, "Mbed Studio: The desktop IDE for Mbed," product homepage, https://os.mbed.com/studio/

27 "JTAG," Wikipedia, last modified October 7, 2024, https://en.wikipedia.org/wiki/JTAG

28 ARM Limited, *Debug Interface v5 Architecture Specification*, 2006.

29 STMicroelectronics. (2024). “ST-LINK/V2 In-circuit Debugger/Programmer for STM8 and STM32,” https://www.st.com/en/development-tools/st-link-v2.1.html?utm_source=platformio&utm_medium=docs
30 QEMU Project Developers, “QEMU: Arm System Emulator,” https://www.qemu.org/docs/master/system/target-arm.html
31 MathWorks, Inc., “Simulink,” product homepage, https://www.sysgo.com/matlab-simulink

9

Server Software Development

9.1 Introduction

This chapter provides an in-depth exploration of the fundamental concepts and technologies that underpin modern web development. It will examine the key components of website development, including the crucial distinction between back-end and front-end development. In the back-end development section, the reader will explore the realm of web servers, analyzing the differences between popular options like Apache and Nginx. The XAMPP server will also be discussed, and the role of databases, including the distinctions between Structured Query Language (SQL) and MySQL, will be investigated. Additionally, the chapter will address MySQL security and vulnerabilities, equipping the reader with the knowledge to build robust and secure database infrastructures.

Shifting the focus to front-end design, the core technologies that enable the creation of dynamic web interfaces, including Hypertext Markup Language (HTML), Cascading Style Sheets (CSS), and JavaScript, will be examined. The rise of front-end frameworks and the use of AJAX for dynamic content presentation will also be explored. Furthermore, the principles of responsive web design will be examined, with a particular emphasis on the popular Bootstrap framework. This will provide the reader with the tools to develop websites that seamlessly adapt to various device sizes and screen resolutions.

Finally, the critical topic of network security will be addressed, including device security, data content security, and operational behavior safety. This will ensure that the reader has a comprehensive understanding of the measures necessary to protect web applications and the data they handle. Practical coding examples will be introduced throughout the chapter in the context of IoT applications. By the end of this chapter, the reader will possess a solid foundation in web development, empowering them to design, build, and secure robust and engaging web experiences.

Introduction to Wireless System Design: From Circuits to Web-based Applications, First Edition.
Henry Lau, Ludy Liu, and Keith Chan.

9.2 Website Development

Website development today involves creating and maintaining websites with modern design and functionality. Key aspects of website development include responsive design, which adapts to different screen sizes and resolutions due to the increasing use of mobile devices. Responsive design ensures an optimal user experience on desktops, tablets, and smartphones. Content Management Systems like WordPress, Drupal, and Joomla make it easier for nontechnical users to create and manage website content. These systems provide user-friendly interfaces, customizable templates, and plugins/extensions to simplify the development and maintenance process.

Website development typically involves two parts: front-end and back-end development. Front-end development focuses on creating the visual and interactive aspects of a website. It involves using HTML, CSS, and JavaScript to create user interfaces, implement designs, and add interactivity. Modern front-end frameworks such as React [1], Angular [2], and Vue.js [3] provide efficient ways to build dynamic and responsive web applications. Back-end development involves programming for web servers and managing databases. Developers use languages such as hypertext preprocessor (PHP), Python, Ruby, or Node.js to handle data processing, server communication, and business logic. Frameworks such as Laravel, Django, and Express.js offer prebuilt components and tools to streamline back-end development.

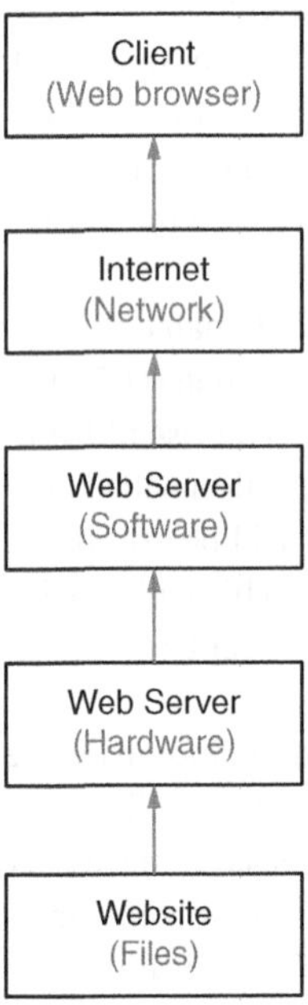

Figure 9.1 Basic components and workflow of a web server

9.3 Back-end Development

9.3.1 Typical Web Servers

A web server is computer software or hardware designed to deliver web pages or other web content to clients, typically web browsers, over the Internet or a local network. The main functions of a web server include hosting websites, processing client requests, managing the Hypertext Transfer Protocol (HTTP), and handling multiple clients simultaneously. The web server stores the files (HTML, CSS, JavaScript, images, etc.) that make up a website and serves them to clients upon request. When a client (typically a web browser) requests a web page, the web server receives the request, processes it, and returns the appropriate web content to the client (see Figure 9.1). Web servers communicate with clients using the HTTP protocol, which is the standard protocol for transmitting data over the web. Additionally, they are designed to handle multiple client requests simultaneously, allowing them to serve content to many users concurrently. Web servers can be deployed on physical hardware, on-premises, or in the cloud, and they can be configured to handle a wide range of web-based applications, from simple static websites to complex, dynamic web applications.

The two main protocols involved in web servers are HTTP and Transmission Control Protocol (TCP). Both protocols are request–response protocols, which means that the client initiates a request, and the server listens to the request and provides a response to the client. Although each web server program is different, there are some common characteristics. Every web server program accepts HTTP requests from the Internet and provides HTTP responses to the requester. HTTP responses typically contain an HTML file, but can also contain a plain text file, an image, or other types of files.

Typically, web servers store these files within their local file systems. The Uniform Resource Locator (URL) and the local file name follow a hierarchical organizational structure. When a request is received, the web server compares the URL to the local file system to locate and retrieve the corresponding file. Once the web server software is successfully installed and configured, the server administrator will designate a local path known as the root directory. This root directory serves as the base location where the server software stores and manages files. The most common web (HTTP) servers on the market today are the Apache Software Foundation's Apache HTTP Server, Microsoft Internet Information Server, Google's Google Web Server, Nginx of Nginx company, and Lighttpd from Lighttpd Inc.

9.3.2 Apache Versus Nginx

Apache is one of the most popular web servers around the world. As of September 2022, Apache holds 31.2% of the market, while Nginx holds about 34.2% of the market [4]. Both servers can establish communication based on the principles of the Representational State Transfer [5] architectural style using Uniform Resource Identifiers (URIs) or URLs. These URIs/URLs serve as addresses to access resources, and the HTTP protocol is utilized to facilitate the delivery of various web services.

Each web server possesses its unique characteristics. Apache, with its extensive development history, offers numerous advantages such as stability, open-source nature, and cross-platform compatibility. Consequently, it is designed as a heavyweight web server. However, Apache lacks support for high concurrency, and concurrent access can lead to excessive memory consumption. Additionally, the process or thread switching performed by the operating system (OS) consumes significant CPU resources, resulting in a decrease in the average response time for HTTP requests. It was during this time that the lightweight, high-concurrency server Nginx emerged as the dominant solution.

Nginx employs an event-driven architecture, allowing it to handle millions of TCP connections. It is a versatile server capable of running on various platforms, including Linux, Windows, FreeBSD, Solaris, and Mac OS. Nginx serves as a free, open-source HTTP server and reverse proxy server. It also functions as an IMAP, POP3, and SMTP proxy server. Furthermore, Nginx can be utilized for website deployment processing and can act as a reverse proxy to implement load balancing.

When performance is a priority, Nginx is the recommended choice for web services. On the other hand, if stability is the primary concern rather than performance, Apache is a suitable option. Apache offers various functional modules that surpass Nginx in certain aspects. For example, Apache excels in its SSL module and provides more configurable options.

9.3.3 XAMPP Server

XAMPP is an open-source software bundle designed for web development. It comprises several key components, including the Apache web server, the MySQL relational database management system (RDBMS), the PHP server-side scripting language, and the Perl programming language. The acronym "XAMPP" stands for "Cross-platform (Windows, macOS, Linux), Apache, MySQL, PHP, and Perl." As a free and easily installable software package, XAMPP provides a complete and self-contained web development environment. This allows developers to create and test their web applications locally, without the need for a live web server or hosting service. The cross-platform nature of XAMPP means it is available for Windows, macOS, and Linux, enabling developers to work on their preferred OS.

One of the primary advantages of XAMPP is its simplicity. It offers a straightforward installation process, making it particularly user-friendly for beginners setting up a local development environment. By including all the necessary components for web development, such as the web server, database, and scripting languages, in a single package, XAMPP streamlines the setup process and allows developers to focus on building their applications. Furthermore, XAMPP features a user-friendly control panel, which enables users to start, stop, and manage the various components of the web development stack. Additionally, the software supports a range of plugins and modules, allowing developers to extend its functionality as needed.

Given its simplicity, cross-platform compatibility, and comprehensive suite of tools, XAMPP has become a widely adopted choice among web developers, particularly those new to the field or those requiring a convenient way to set up a local development environment for their projects.

9.3.4 Database

A computer database is a structured collection of data that is organized and stored electronically for efficient access, management, and retrieval. It is designed to provide a centralized and organized way to store and manage large amounts of information. Some key characteristics of a computer database include its use of a specific data model, such as the relational model, to organize data into tables, rows, and columns, which allows for efficient storage and querying of the data. Databases also enforce rules and constraints to ensure the accuracy, consistency, and completeness of the stored data, helping to maintain its integrity. Additionally, they often include security features, such as user authentication, access control, and backup and recovery mechanisms, to protect the data from unauthorized access, modification, or loss.

Databases provide a query language, typically SQL, that allows users to retrieve, filter, and manipulate data based on specific criteria. They also manage concurrent access to data, ensuring that multiple users or processes can access and modify the data simultaneously without causing conflicts or data inconsistencies. Furthermore, databases are designed to be scalable, able to handle large amounts of data and support a growing number of users and applications without significant performance degradation. They are used in a wide range of applications, from small personal databases to large-scale enterprise systems, and are an integral part of many software applications, including e-commerce platforms, customer relationship management systems, and financial management systems.

9.3.5 Database Infrastructure

A simplified SQL database architecture for IoT applications is shown in Figure 9.2.

As can be seen from Figure 9.2, when a client operates an app or an IoT sensor (detection device) to collect measurement data from the environment, the data generated is sent over the network to a database management system (DBMS) [6] and stored in a database. The user retrieves the data from the database through SQL queries. A DBMS is large-scale software used to manage the database. Representative DBMS include Oracle, Microsoft SQL Server, MySQL, and PostgreSQL. It can assist us in adding, reading, updating, and deleting data. Without a DBMS, it is impossible to manipulate data at scale.

There are two types of DBMS: relational and nonrelational, also known as SQL and NoSQL, respectively. A relational database is a type of database that stores organized data in tables related to one another.

In the 1970s, E.F. Codd first proposed the relational data model. IBM then developed the earliest SQL language in the System RDBMS. In October 1986, ANSI (American National Standards Institute) adopted SQL as the standard language for RDBMS.

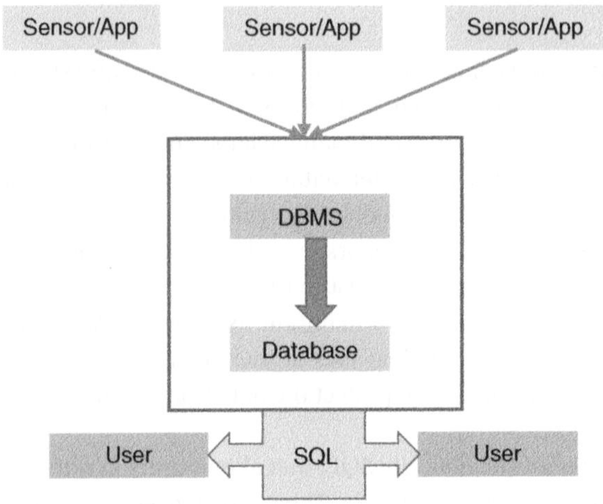

Figure 9.2 Simplified SQL database architecture for IOT application

Service quality improvement, productivity increases, efficiency management, and decision-making assistance are among the expected benefits of big data for enterprises. The effective retrieval of useful information from databases, particularly when dealing with large volumes of data originating from tens of thousands of devices, is a pressing concern for businesses. The availability of a tool that can swiftly locate valuable information in the database – namely SQL – is a shared aspiration. In addition to SQL, there are other tools and techniques that can be used to extract valuable information from databases, such as NoSQL and others.

9.3.6 Difference Between SQL and MySQL

SQL is a programming language used to manage RDBMS. MySQL is one of these systems; that is, it is database software. MySQL was originally developed by MySQL AB in Sweden and was later acquired by Oracle in 2010. Since MySQL is an implementation of the SQL standard, the relationship between MySQL and SQL is similar to the difference between C++ and VC++. One is the language, while the other is the software implemented based on that language, which forms the link between the two.

Users can write SQL queries to retrieve, add, or manipulate data based on standard SQL syntax. Of course, they can also use other database management software, such as Microsoft's SQL Server, Oracle, or NoSQL databases. If a sufficient budget is available to the company, they can choose expensive hardware and enterprise software, creating an ideal work environment.

For small and medium-sized enterprises with limited resources, the primary consideration is cost. Since MySQL is open source and the Community Edition [7] is available for free, it is ideal for companies that do not have high visitor traffic. MySQL offers a simple setup process, ease of use, and compatibility with all major OS. Furthermore, MySQL is an integral part of WAMP/LAMP, a widely used open-source software stack for web applications, which stands for Windows/Linux, Apache, MySQL, and PHP/Perl/Python.

9.3.7 MySQL Security Improvement

While MySQL is a relatively easy-to-use database management system, proper security precautions must be taken to prevent it from being hacked. First and foremost, all anonymous accounts need to be removed. Additionally, a strong password must be set for the root account, as the use of an empty password is unacceptable. It is also important to avoid the use of default or weak passwords. The default MySQL installation process creates a "test" database, which may contain sample data, scripts, or configurations that could pose a security risk. It is highly recommended that this "test" database be removed. Since it is possible to create multiple user accounts on a MySQL server, it is advisable that MySQL be run with different users. Furthermore, user permissions in the MySQL database need to be properly managed. If remote administration is not required, the remote connection to the database should be disabled.

MySQL stores the commands typed in the client history file, which may contain sensitive information such as passwords. If it is not necessary, the MySQL command history should be disabled. Additionally, MySQL can communicate with the local file system, so it is advisable that MySQL be prevented from accessing local files to avoid potential malicious attacks. The MySQL configuration file, called "my.ini" on Windows and "my.cnf" on Linux, is crucial for configuring various settings and parameters of the MySQL server. The file permissions on this configuration file need to be set appropriately.

Regular database backups are essential for protecting and restoring databases. It is, therefore, crucial that databases be backed up on a regular basis.

9.3.8 Some Web Application Vulnerabilities

9.3.8.1 SQL Injection

SQL injection is one of the most common attacks used by hackers to steal information or cause data loss by targeting databases. SQL injection [8] is an injection-type attack in which hackers run malicious SQL queries that can lead to serious consequences, such as data loss or theft. To prevent SQL injection, it is recommended to use prepared statements or parameterized queries. Prepared statements separate the SQL code from the data, eliminating the need for manual escaping or stripping slashes.

9.3.8.2 Parameterized Query (Prepared Statement)

Here is an example of using prepared statements in PHP with the MySQLi extension. The code uses prepared statements to insert data into a table named "users" with columns "username" and "password":

```
$stmt = $conn->prepare ("INSERT INTO users (username,
password) VALUES(?, ?)");
```

The `prepare()` method is used to prepare the SQL query with placeholders (?) for the values that will be inserted.

```
// Bind the user input to the placeholder
$stmt->bind_param("ss", $username, $password);
```

The `bind_param()` method is then used to bind the actual values to the placeholders. In this case, the placeholders are specified as "ss," indicating that both the username and password values are strings. This approach prevents SQL injection by separating the SQL code from the user input.

9.3.8.3 String Escaping

This approach can be useful for reducing the risk of SQL injection in scenarios where parameterized queries are unavailable or impractical. An example of a PHP script is as follows:

```
//to prevent mysqli injection
// escapes special characters in a string for use in an SQL
query
$username = mysqli_real_escape_string($con, $username);
$password = mysqli_real_escape_string($con, $password);
```

The `mysqli_real_escape_string()` function is a PHP function designed to escape special characters in a string before sending a query to MySQL. The function adds a backslash to special characters in the input string. For example, the attempted attack string `" ' OR ' '=' "` is sanitized to `" \' OR \'\'=\' "`. This function is an effective method for mitigating the risk of SQL injection when using the MySQLi extension in PHP. Many developers utilize this function to escape single quotes in strings, thereby helping to prevent SQL injection attacks.

9.3.8.4 DDoS Attacks

One common brute-force attack against MySQL databases is a distributed denial-of-service (DDoS) attack. Attackers use multiple accounts to send a series of fraudulent queries to inundate the database in quick succession. The purpose of this attack is to slow down the database and eventually crash it. To mitigate DDoS attacks, several strategies can be employed, including limiting the maximum number of connections, disabling persistent connections, and closing open connections that cause errors.

9.3.8.5 Cross-Site Scripting (XSS)

Cross-site scripting attacks (XSS) [9] represent a significant danger to web applications, exploiting website vulnerabilities through the injection of malicious scripts that execute on the client side. This attack is typically performed within the browser, aiming to embed and execute JavaScript in a web page of the targeted application, granting hackers control over the current page and compromising the application's security. The `mysql_real_escape_string()` function mentioned earlier does not prevent XSS; it only makes SQL injection impossible.

To combat XSS, it is advisable to utilize the functions `htmlspecialchars()` or `strip_tags()`. The `htmlspecialchars()` function converts special characters, such as "<", into their corresponding HTML entities (e.g., "<"), ensuring they are displayed as intended on the page without being executed as part of a script. On the other hand, the `strip_tags()` function removes all HTML tags from the input, effectively mitigating the risk of XSS by eliminating any potential script tags or malicious code. It is recommended to use the standard `mysql_real_escape_string()` function when inputting data into the database. Additionally, when querying content from the database, the `strip_tags()` function should be employed before outputting the data to the screen.

An example of a PHP script is as follows:

```
echo strip_tags("<script>alert('HACKED')</script>");
```

The corresponding output will be:

```
alert('HACKED')
```

This demonstrates that no JavaScript has been executed.

9.3.9 Limitations

MySQL is a database management system that is highly versatile and can be scaled effectively in accordance with the needs of the business. It is suitable for projects of varying sizes and can handle daily transaction volumes of up to millions. However, it's important to note that MySQL may not be the ideal choice for environments that require a robust memory-focused search engine, as it is not specifically designed for extremely high-concurrency scenarios. Additionally, MySQL is associated with relatively higher overhead and may not provide optimal speed in certain circumstances.

9.4 Front-end Design

9.4.1 Building Dynamic Web Interface by HTML, CSS, and JavaScript

This section introduces building web pages for IoT applications through a few examples that utilize simple, basic, and commonly used HTML elements. Web pages typically contain text, images, and other information that must be presented to users in an accurate and visually appealing manner. This is the responsibility of web page editors, who work to ensure a satisfying user experience. At the core of dynamic web interfaces, three key technologies that work in harmony: HTML, CSS, and JavaScript.

HTML is the markup language used to structure and define the content of web pages. It provides the basic building blocks for presenting information to users. CSS is then employed to control the visual appearance and layout of these HTML elements, allowing designers to customize the fonts, colors, spacing, and overall aesthetics of the web page. Finally, JavaScript is a dynamic client-side scripting language that enhances the interactivity and dynamic behavior of web applications. It can be used to add animations, respond to user actions, and create rich, engaging interfaces.

Being a crucial component underpinning web technologies, the Document Object Model (DOM) is a programming interface for web documents that defines the logical structure of a web page and how it can be accessed and manipulated by programs and scripting languages like JavaScript. Specifically, the DOM represents the web page as a tree-like structure, with the initial HTML document serving as the root (see Figure 9.3). Each HTML element, such as `<div>`, `<p>`, or `<img>`, is considered a node in this tree structure, and the text content and attributes of these elements are also represented as nodes.

To demonstrate these concepts, Code 9.1 presents a simple example web page format. In this example, the `<div>` element is commonly used by developers to create sections or divisions within a web page layout, apply CSS styles, and manipulate content using JavaScript. It serves as a fundamental building block for creating responsive and flexible web designs.

9.4.2 Cascading Style Sheets

CSS is a styling language that enhances the presentation and appearance of HTML documents. It enables web developers to apply various styles, such as colors, fonts, layouts, and animations, to elements within a web page. CSS works by selecting

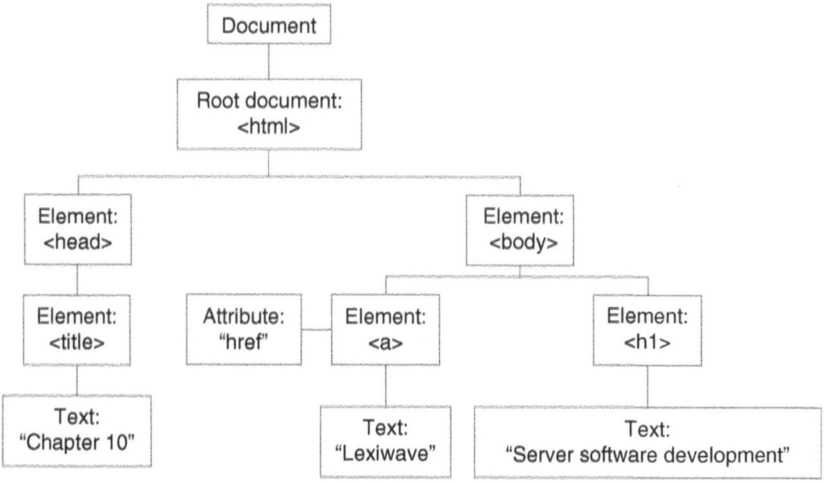

Figure 9.3 HTML DOM tree

Code 9.1 Web page Protoype Code

````
```html
<!DOCTYPE html>
<html>
 <head>
  <title>Chapter 10</title>
 </head>
 <body>
  <h1>Server Software Development</h1>
  <a href="http://lexiwave.com">Lexiwave</a>
 </body>
</html>
```
````

HTML elements and applying specific rules that define how those elements should be displayed on the screen. This allows for consistent styling across multiple web pages, simplifies the process of updating the visual design, and provides flexibility in creating visually appealing and user-friendly web interfaces.

Code 9.2 Code Snippet of Internal Style Sheet

````
```html
<!DOCTYPE html>
<html>
<head>
<style>
selector {
    CSS property: value;
}
</style>
</head>
```
````

There are three main ways to set CSS: inline style (writing CSS directly within HTML tags), internal style sheet (writing CSS in the file header), and external style sheet (creating an external CSS file with the .css extension). An example of an internal CSS style sheet is shown in Code 9.2.

Two commonly used attributes in HTML elements that allow CSS or JavaScript to locate and manipulate DOM elements are `id` and `class`. The `id` attribute should only be used once per HTML element on a web page, while the `class` attribute can be applied to multiple HTML elements. In CSS, the `id` selector is denoted by `#`, and the `class` selector is denoted by `.`. The JavaScript function `getElementById` can only be used with the `id` selector, not with the `class` selector. An example of using `id` and `class` to set the background color is shown in Code 9.3.

In this example, the `.cl-1` class applies a light gray background color and blue text to the `<h2>` and `<p>` elements, while the `#id-1` ID applies a blue-gray background color and white text to the `<h1>` and `<div>` elements as shown in Figure 9.4.

9.4.3 JavaScript

JavaScript was first introduced in 1995 as LiveScript and was later renamed. Netscape released JavaScript Version 1.0 in 1996. Netscape Navigator, one of the earliest web browsers, is well-known to us.

JavaScript is commonly used in modern web pages, but it can also be embedded in PDFs to create small desktop plugins and even serve as the foundation for dynamic server-side functionality. In the past, JavaScript applications were known

Code 9.3 Code Snippet of Using `id` and `class`

````
```html
<style>
    .cl-1 {
        background-color: #EBEBEB;
        color: blue;
    }
    #id-1 {
        background-color: #ABBAEA;
        color: white;
    }
</style>
<body>
    <h1 id="id-1"> Have a nice day.</h1>
    <h2 class="cl-1"> Have a nice day. </h2>
    <p class="cl-1"> Have a nice day. </p>
    <div id="id-1"> Have a nice day. </div>
</body>
```
````

Figure 9.4 Result of using `id` and `class` attributes

for being visually unappealing and were often used for intrusive purposes such as pop-up windows, screen vibrations, and audio playback. They were also used for common widgets such as image flipping and browser navigation bars, although with less annoyance. JavaScript was most popular in the 1990s. One of its notable applications is creating and validating HTML forms. However, early adoption of JavaScript was limited due to slow broadband speeds and challenges in writing cross-browser JavaScript, which hindered its widespread use and development.

JavaScript has experienced a resurgence in the past 15 years due to three significant developments since its inception. Firstly, the emergence of AJAX technology allowed for seamless data retrieval and asynchronous communication between the browser and server, enhancing the interactivity of web applications. Secondly, browsers underwent substantial improvements, implementing advanced features and better performance, resulting in a more robust JavaScript execution environment. Lastly, the development of frameworks has provided developers with powerful tools and libraries that have expanded their capabilities and simplified the development process of JavaScript applications. These advancements have significantly broadened the functionality and user-friendliness of JavaScript.

9.4.4 The Rise of Frameworks

A JavaScript framework [10] is a library of code designed to speed up the development process. The advantage of using a framework is that it provides powerful and flexible functions to enhance the web interface and its functionality, with the most important benefit being the establishment of cross-browser reliability. However, the disadvantage is that learning the framework can be more difficult, requiring advanced customization, and using the framework necessitates downloading it rather than using JavaScript alone. More code in JavaScript can lead to decreased performance.

In the past, jQuery was a popular framework that played an important role in web development. It served as a feature-rich JavaScript library primarily used for manipulating DOM elements. Additionally, the core functionality of jQuery enhanced asynchronous transfer (AJAX) and event handling, simplifying the management of remote files and events. Within jQuery, the global `$` variable is assigned to an object that has various unique behaviors. As a result, variables beginning with `$` are often reserved for jQuery-related values.

While jQuery has played a significant role in DOM manipulation, it does not follow the general trend of technological advancement and is expected to be eventually replaced. However, it remains a powerful tool for beginners in front-end development. Currently, the three prominent front-end frameworks are Vue, React, and Angular. PHP is still widely used by back-end programmers. A major trend in development is to write back-end code using PHP or Node.js, which eliminates the reliance on server-generated HTML code.

9.4.5 Dynamic Presentation Using AJAX

9.4.5.1 Asynchronous JavaScript and Extensible Markup Language (AJAX)

AJAX [11], short for Asynchronous JavaScript and Extensible Markup Language (XML), is a web development technique that allows asynchronous communication between client-side and server-side scripts. It uses a nonstandard XMLHttpRequest [12] object to facilitate data exchange. Ajax can send and receive information in various formats, including XML, HTML, and text files. One of the main advantages of Ajax is its ability to perform actions without refreshing the entire page, thanks to its asynchronous nature. This means that only a specific portion of the page can be updated based on user events, resulting in a more seamless and dynamic user experience. AJAX requests can be made using either the native JavaScript XMLHttpRequest or jQuery.

9.4.5.2 JavaScript Object Notation (JSON)

JavaScript Object Notation (JSON) [13] is a widely used data exchange format that facilitates the transfer of data between a server and a web application. Prior to JSON, data exchange typically involved XML files or text files (TXT), which often required dealing with individual files, making it cumbersome to open, read, update, and write data. With the introduction of JSON, these limitations were overcome. JSON quickly gained popularity because it was specifically designed for efficient data exchange and provides faster performance due to its concise data encoding. JSON supports multiple data types, including strings, numbers, Booleans (true or false), arrays, and objects. The data structure in JSON is enclosed in curly braces `{}` to define an object. Figure 9.5 shows a typical structure of JSON data along with some reference data.

(a)

~~~~
---------------------------------------------------------------------
```json
{
    "key1": "value1",
    "key2": "value2",
    "key3": ["item1", "item2", "item3"],
    "key4": {
        "nestedKey": "nestedValue"
    }
}
```
---------------------------------------------------------------
~~~~

(b)


```
---------------------------------------------------------------
[{"ID":"2022-09-01 11:38:57","Temp":"77"},{"ID":"2022-09-01
11:38:58","Temp":"91"},{"ID":"2022-09-01 11:38:59","Temp":"94"},{"ID":"2022-
09-01 11:39:00","Temp":"58"},{"ID":"2022-09-01
11:39:01","Temp":"58"},{"ID":"2022-09-01 11:39:02","Temp":"65"},{"ID":"2022-
09-01 11:39:03","Temp":"75"},{"ID":"2022-09-01
11:39:04","Temp":"96"},{"ID":"2022-09-01 11:39:05","Temp":"41"},{"ID":"2022-
09-01 11:39:06","Temp":"65"}]
---------------------------------------------------------------
```


Figure 9.5 (a) Typical structure of JSON data (b) JSON data

The given JSON data represents a collection of temperature measurements taken at different timestamps. Each measurement object has two properties: "ID" and "Temp." The ID property represents the timestamp when the temperature reading was taken, and the Temp property represents the corresponding temperature value at that specific time. The JSON data provides a time-series record of temperature readings, enabling the analysis and interpretation of temperature trends over time. Due to the advantages of the JSON format, many developers use JSON with AJAX to pass data between the client and server. A website updating sensor data can be considered an example of AJAX usage. If this data is used on the website, it must be stored on the server so that it can be retrieved when needed.

9.4.5.3 AJAX Call in jQuery

To implement an AJAX call every minute using JavaScript, the AJAX logic should be placed inside a function, and the `` `setInterval()` `` function should be used to create the recurring interval. A sample code snippet is shown in Code 9.4.

Code 9.4 Code Snippet of AJAX Call

```
// Define the function that performs the AJAX call
function ajaxCall() {
   $.ajax({
      url: 'your-URL-here',
      type: 'POST',
      data: { key: 'value' },
      success: function(response) {
        // Perform your AJAX operations in this section.
         console.log(response);
      },
      error: function(xhr, status, error) {
         // Handle the error if the AJAX request fails
         console.error(error);
      }
   });
}
// Set up the interval to call the function every
1 minute (60000 milliseconds)
setInterval(ajaxCall, 60000);
```

Code 9.5 Code Snippet of Web page Making Ajax Call

```
Index.php
<script src="http://ajax.googleapis.com/ajax/libs/
jquery/3.6.0/jquery.min.js"></script>
<body>
<script>
$(document).ready(function() {
// Wait till page is loaded
setInterval(ajax_Call, 3000);

function ajax_Call() {
$('#my-id').load('random.php', function() {
//      });
   }
});
</script>

<h1>Ajax Div refresh demo</h1>
<div id="my-id"></div>
</body>
```

In this code snippet, the `` `ajaxCall()` `` function represents the logic for the AJAX operations. The `` `setInterval()` `` function is used to schedule the execution of the `` `ajaxCall()` `` function every 1 minute (60000 ms).

The AJAX call feature can be extended to update a content division of a web page automatically as shown in Code 9.5. In this example, the page "random.php" will be loaded [14] into a `` `<div>` `` element with the ID "my-id" every 3 seconds. This script allows for the dynamic display of a random lucky number between 10 and 100 each time the script is executed.

The code sets up a simple AJAX-based auto-refreshing mechanism. Every 3 seconds, the JavaScript code makes an AJAX request to the `` `random.php` `` file as shown in Code 9.6, which generates a random number between 10 and 100, and returns it as an HTML snippet. The JavaScript code then updates the content of the `` `<div>` `` element with the ID `` `my-id` `` with the new random number. The result is shown in Figure 9.6.

Code 9.6 Code Snippet of Getting Random Number

```
random.php
<?php
$string1='Lucky number :';
$num1=rand(10,100);
echo '<h2>';
echo $string1 . $num1;
echo '</h2>';
?>
```

Figure 9.6 Result on content update in document division

This can be useful for creating dynamic web pages that need to update their content regularly without requiring a full page refresh, such as a live counter or a randomly generated element.

9.5 Responsive Web Design Using Bootstrap

Today, a multitude of devices are employed to gain access to the web, including tablets with smaller screens and mobile phones with a common resolution of around 720px. In contrast, iPads boast a resolution of approximately 2048px, while widescreen computers offer a resolution of 1920px. Therefore, it is crucial to ensure that content appears appealing on screens of different sizes.

Responsive Web Design (RWD) is an approach that aims to create websites that are easily readable and navigable across various devices. It eliminates the need for users to zoom in, pan, scroll extensively, or engage in other cumbersome actions on small-device browsers. RWD utilizes HTML5 and CSS3 media queries [15] to specify page width and apply different CSS styles for different screen widths. Separate CSS files are created for each screen width to ensure optimal rendering.

By adopting RWD principles, web developers can maintain accessibility and visual appeal across different screen sizes and browsing devices.

9.5.1 Bootstrap

Bootstrap is a popular and widely used front-end framework for web development, helping create responsive, mobile-first websites and web applications. It was initially developed by Twitter and is now maintained by a community of developers. The framework's key aspects in web design include a responsive design with a grid system that automatically rearranges and resizes content based on the screen size, a wide range of predesigned CSS and JavaScript components that can be easily customized and integrated, and a comprehensive set of utility classes that allow for quick application of common styles and layout properties. While a solid foundation is provided, extensive customization is also allowed, enabling developers to override default styles, create custom CSS, and integrate bespoke JavaScript functionality. Importantly, the framework is designed to be compatible across all modern web browsers, ensuring a consistent look and functionality. A mobile-first approach is also encouraged, where the website is optimized for small screens first and then enhanced for larger displays. Additionally, accessibility-focused features, such as proper semantic markup, Accessible Rich Internet Applications (ARIA) attributes, and keyboard-friendly interactions, are included to improve inclusivity.

Several key benefits are offered by using Bootstrap, such as rapid prototyping through its prebuilt components and utilities, a truly responsive and mobile-friendly design, and a consistent look and feel across web pages. Furthermore, the framework's large and active community, along with its extensive documentation, makes it easy for developers to get started and find solutions to common problems.

The Bootstrap grid [16] system is a core feature that facilitates responsive web design. Based on a 12-column layout, the grid system allows for the creation of complex and responsive page structures. This is achieved through the use of rows to create horizontal groups of columns that contain the web page content. Each row can have up to 12 columns, with the width of each column specified as a fraction of the 12-column grid. Additionally, several responsive breakpoints based on screen size are defined, enabling the specification of different column widths for each. Figure 9.7 illustrates the effect of the grid system for a responsive web page, where the screen changes from a full-screen span of one row when viewed on a large-screen device to stacked rows when the screen width is less than 992px.

Another web page utilizing responsive web design, featuring a header and two charts, is shown in Figure 9.8. It is a simple dashboard using Bootstrap 5 and Chart.js to present sensor data in a visually appealing way. The two charts will stack together when the screen size is smaller.

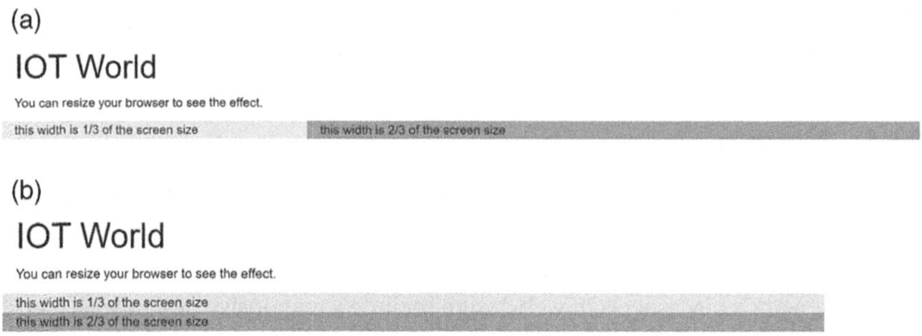

Figure 9.7 (a) Screen viewed on a large-screen device and (b) stacked columns viewed on a screen with smaller width

9.6 Network Security

In the digital landscape, data security is of paramount importance for enterprises, especially due to the ease with which online attacks can occur. Cybersecurity involves safeguarding networked systems, including hardware, software, and data, against various cyber threats. The field of cybersecurity can be broadly categorized into three sections, each serving a distinct purpose. These include Device system security, Data Content Security, and Operational Behavior Safety.

9.6.1 Device Security

Device security is essential for safeguarding physical devices used for data storage, computation, and other functions, including servers and hard drives. It forms the foundation of network security and plays a vital role in overall system security. Without proper device security, all other security measures become ineffective, as the compromised or damaged device undermines their effectiveness. Therefore, ensuring the protection and integrity of devices is critical in maintaining a secure network environment. When considering the security of equipment systems, factors such as availability, reliability, and stability should be taken into account. For example, for personal computers, it is crucial to keep the system version up to date, including the OS and applications, and to have reliable antivirus software and a robust firewall in place.

9.6.2 Data Content Security

The term "content security" refers to the encryption of the data itself and the security of the transmission. There are two main encryption methods in use:

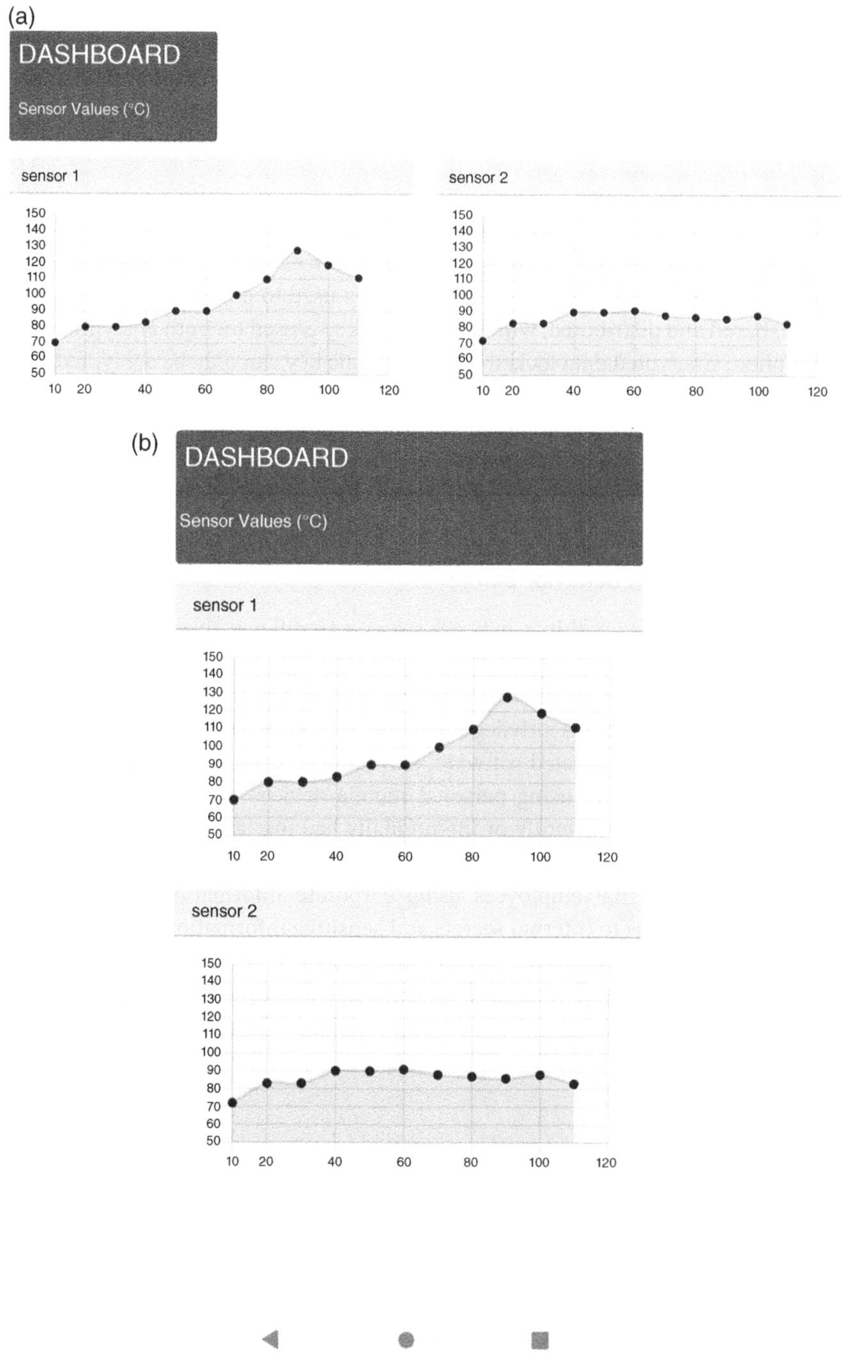

Figure 9.8 (a) Screen viewed on a large-screen device and (b) stacked charts viewed on a smaller screen

9.6.2.1 Symmetric Encryption (Private Key Encryption)

Private key encryption is the type of encryption where only a single private key can encrypt and decrypt information. The advantages are that both the encryption and decryption efficiency and the encryption strength are very high. However, when the key is sent, there is a significant risk that it will be intercepted by hackers.

9.6.2.2 Asymmetric Encryption (Public Key Encryption)

Asymmetric encryption is a cryptographic algorithm based on a pair of keys: a public key and a private key. The public key [17] is used to encrypt data and can be freely shared and distributed, while the private key is used for both encryption and decryption. Encrypted data locked with the public key can only be decrypted with the corresponding private key. These two keys are mathematically linked, but it is computationally infeasible to derive the private key from the public key alone. In the absence of both keys, unauthorized access to the encrypted data is essentially futile, as it remains unintelligible and useless to potential hackers.

9.6.3 Operational Behavior Safety

Operational behavior within a network poses a significant threat to its security. A study conducted by IBM found that a staggering 95% of cybersecurity breaches are the result of human error [18]. Such errors often result from ineffective management practices and lax privilege management systems, leading to activities such as downloading infected software, maintaining weak passwords, disclosing IP addresses, employees using personal mobile devices to transfer data across the network, and inadvertently or intentionally leaking large amounts of data.

With social engineering threats, criminals take advantage of individuals' trust by impersonating internal employees using corporate information. They deceive others to gain access to internal secrets and sensitive information, compromising the security of the network. To mitigate these risks, it is essential to implement robust management practices and strong privilege management systems that promote awareness and best practices to prevent human errors that can lead to cybersecurity breaches.

References

1 "React JS Development," last modified February 29, 2024, https://radixweb.com/react-js-introduction
2 "Best AngularJS Frameworks to Build Impressive Frontend Apps," last modified February 29, 2024, https://www.elluminatiinc.com/best-angulars-frameworks/

3 "Introduction," last modified February 29, 2024, https://vuejs.org/guide/introduction.html

4 "Top Web Server 2022," last modified February 5, 2024, https://www.stackscale.com/blog/top-web-servers/

5 "What is REST?" last modified February 5, 2024, https://www.codecademy.com/article/what-is-rest

6 Silberschatz, A., Korth, H.F., and Sudarshan, S. (2010). *Database System Concepts*, 6e, 3. TBS.

7 "MySQL Community Edition," last modified February 5, 2024, https://www.mysql.com/products/community/

8 Clarke, J. (2012). *SQL Injection Attacks and Defense*, 2e. Syngress.

9 Stuttard, D. and Pinto, M. (2011). *The Web Application Hacker's Handbook: Finding and Exploiting Security Flaws*, 2e, chap. 10. Wiley.

10 Toal, R., "What is JavaScript Framework?" last modified February 5, 2024, https://codeinstitute.net/global/blog/javascript-framework/

11 Holdener, A.T. III (2008). *Ajax: The Definitive Guide.* O'Reilly Media.

12 "XMLHttpRequest," last modified February 5, 2024, https://developer.mozilla.org/en-US/docs/Web/API/XMLHttpRequest

13 "Introducing JSON," last modified February 5, 2024, https://www.json.org/json-en.html

14 "jQuery - AJAX load() Method," last modified February 5, 2024, https://www.w3schools.com/jquery/jquery_ajax_load.asp

15 "CSS Media Queries," last modified February 5, 2024, https://www.w3schools.com/css/css3_mediaqueries.asp

16 "Grid System," last modified February 5, 2024, https://getbootstrap.com/docs/4.0/layout/grid/

17 Stallings, W. (2024). *Cryptography and Network Security: Principles and Practice*, 8e. Pearson.

18 "The Role of Human Error in Cybersecurity Breach," last modified February 5, 2024, https://www.engineeringnews.co.za/article/the-role-of-human-error-in-cybersecurity-breach-2022-08-29

10

Smartphone Application Software Development

10.1 Introduction

A mobile application, commonly called an app, is a piece of application software designed to run on a mobile device such as a smartphone or tablet. Mobile applications often provide users with services similar to those accessed on personal computers. Mobile devices can be considered the end-user interfaces for a wide range of wireless systems and applications. Advancements in areas like big data, mobility, and cloud services have enabled the development of various wireless applications and services. As a result, mobile applications have become an important component of modern information technology.

This chapter serves as a comprehensive guide to wireless system application development, focusing on Internet of Things (IoT) applications. It covers the tools, frameworks, and development environments used to create applications for different mobile platforms, including Android, iOS, and cross-platform solutions. Additionally, it explores the integration of mobile applications with external services and systems, providing readers with valuable insights and practical knowledge to develop versatile and feature-rich IoT applications.

10.2 Smartphone Application Development

Developing modern smartphone applications involves creating and implementing software designed specifically for mobile devices. This process encompasses planning, designing, coding, testing, and deploying applications for platforms such as iOS and Android. Developers utilize programming languages like Swift, Objective-C, Java, and Kotlin, along with various frameworks and tools, to build user-friendly and feature-rich applications. The development process often includes

Introduction to Wireless System Design: From Circuits to Web-based Applications, First Edition.
Henry Lau, Ludy Liu, and Keith Chan.

integration with device hardware, leveraging application programming interfaces (APIs), and optimizing performance for diverse screen sizes and resolutions.

10.3 Tools for Application Software Development

Commonly used tools include Integrated Development Environments (IDEs) like Xcode for iOS and Android Studio for Android. These IDEs provide developers with a comprehensive set of tools, including code editors, debuggers, emulators, and project management features. These IDEs offer a convenient and efficient environment for writing, testing, and debugging mobile applications. Software Development Kits (SDKs) offer a collection of software tools and libraries that enable developers to build applications for specific platforms. For instance, the iOS SDK includes frameworks such as UIKit and Core Data, while the Android SDK includes libraries like Android Support Library and Google Play Services. SDKs simplify the development process by providing pre-built components and APIs.

Cross-platform development frameworks, such as React Native, Flutter, and Xamarin, allow developers to write code once and deploy it on multiple platforms, such as iOS and Android. These frameworks leverage a single codebase and provide access to native features and functionality, offering time and effort savings for developers targeting multiple platforms. In the following sections, simple apps will be introduced and developed using Android Studio, React Native, Flutter, and Swift to provide readers with practical knowledge on writing apps.

10.4 Android Studio

Android Studio [1] was launched by Google as an IDE for Android. It aims to provide a development environment that introduces new tools for building Android apps. When a new project is created in Android Studio, almost all files in the SDK directory can be seen in the project structure. The Gradle-based management system [2] in an Android project is a build system that helps automate and manage the project's build process. It is responsible for compiling the source code, managing dependencies, and generating the Android Application Package (APK) file for deployment. The "build.gradle" file, written in Groovy or Kotlin, is used to define the project's configuration and build tasks. It allows developers to specify dependencies, define custom build logic, and configure various aspects of the project, such as merging resources, signing the APK, and generating different build variants. The Gradle-based management system has also been further improved, making the development process more flexible.

10.4.1 Android Programming on Simple TCP Client

In industry and family life, the IoT is widely utilized today, encompassing the remote control of industrial equipment, smart home appliances, and smart switches through computers and mobile phones. When a mobile phone is connected to an IoT system, it transforms into a client that sends data to or retrieves information from the host. Consequently, when developing mobile apps using various tools, the transmission control protocol (TCP) Client application is recommended as the primary choice for illustrating the programming of the interaction between mobile devices and IoT hosts.

The following is the complete `MainActivity.java` program written in Java using Android Studio. Android developers can create Android applications using the Java and Kotlin [3] languages.

10.4.2 Explanation of the Code

The code shown in Code 10.1 is an Android application that serves as a TCP client. It allows the user to establish a TCP connection with a server, send a message, and receive a response from the server. A simple description of the code follows:

The `MainActivity` class extends the `AppCompatActivity` class, which is the base class for activities that use the support library's features. Several classes are imported by the program to create its functionality. For example, the `java.net.Socket` class allows us to create socket objects, which help implement all basic socket operations. The `OutputStream` class is an abstract class that enables writing data to an output sink or destination, such as a socket.

Code 10.1 Android Code for a TCP Client

MainActivity.java

```
package com.example.tcp_client3;
import androidx.appcompat.app.AppCompatActivity;
import android.content.SharedPreferences;
import android.os.AsyncTask;
import android.os.Bundle;
import android.util.Log;
import android.view.View;
import android.widget.EditText;
import android.widget.TextView;
import java.io.BufferedReader;
```

```
import java.io.IOException;
import java.io.InputStreamReader;
import java.io.OutputStream;
import java.io.PrintWriter;
import java.net.InetSocketAddress;
import java.net.Socket;
import java.net.SocketAddress;
import java.net.SocketException;

public class MainActivity extends AppCompatActivity {

    public static String server_address = "xxx.xxx.xxx.
xxx";
    public static int server_port = 10000;
    private LongOperation lo = null;

    public static final String MY_PREFS_NAME =
"MyPrefsFile";

    @Override
    protected void onCreate(Bundle savedInstanceState) {
        super.onCreate(savedInstanceState);
        setContentView(R.layout.activity_main);

        SharedPreferences prefs = getSharedPreferences
(MY_PREFS_NAME, MODE_PRIVATE);
        String nserver_address = prefs.getString
("server", null);
        int nserver_port = prefs.getInt("port", 10000);

        TextView server_text = findViewById(R.id.
serverText);
        server_text.setText(nserver_address != null ?
nserver_address : "");

       TextView port_text = findViewById(R.id.portText);
       port_text.setText(nserver_port != 10000 ? String.
valueOf(nserver_port) : "");
```

```
        TextView textView = findViewById(R.id.textView2);
        textView.setText("No Response Yet\nTry Sending
 Some Data.");

        EditText editText = findViewById(R.id.editText);
        editText.setText("Message");
    }

    private class LongOperation extends
AsyncTask<String, Void, String> {

        private Socket socket;

        @Override
        protected String doInBackground(String... params)
{
            socket = null;
            SocketAddress address = new InetSocketAddress
(server_address, server_port);

            socket = new Socket();

            try {
                socket.connect(address, 3000);
            } catch (IOException e) {
                Log.d("time", "idle");
                e.printStackTrace();
            }

            try {
                socket.setSoTimeout(3000);
            } catch (SocketException e) {
                Log.d("timeout", "Server took too long
                to respond");
                e.printStackTrace();
                return "Can't Connect";
            }
```

```
        OutputStream out = null;
        try {
            out = socket.getOutputStream();
        } catch (IOException e) {
            e.printStackTrace();
        }
        PrintWriter output = new PrintWriter(out);

        output.print(params[0]);
        output.flush();

        String str = "waiting";
        BufferedReader br = null;
        try {
            br = new BufferedReader(new
            InputStreamReader(socket.
            getInputStream()));
        } catch (IOException e) {
            e.printStackTrace();
        }
        try {
            String line;
            str = "";
            while ((line = br.readLine()) != null) {
                str = str + line;
                str = str + "\r\n";
            }
        } catch (IOException e) {
            e.printStackTrace();
        }

        if (str != null) {
            Log.d("test", "Print what you just
            read.");
            System.out.println(str);
        }
```

```
            output.close();
            try {
                br.close();
            } catch (IOException e) {
                e.printStackTrace();
            }
            try {
                socket.close();
            } catch (IOException e) {
                e.printStackTrace();
            }
            Log.d("tag", "done server");
            return str;
        }

        @Override
        protected void onPostExecute(String result) {
            TextView textView = findViewById(R.id.
textView2);
            textView.setText(result);
        }

        @Override
        protected void onPreExecute() {
            TextView server_text = findViewById(R.id.
serverText);
            TextView port_text = findViewById(R.id.
portText);

            server_address = server_text.getText().
toString();
            server_port = Integer.parseInt(port_text.
getText().toString());

            SharedPreferences.Editor editor =
getSharedPreferences(MY_PREFS_NAME, MODE_PRIVATE).
edit();
```

```
            editor.putString("server", server_address);
            editor.putInt("port", server_port);
            editor.apply();
        }

        @Override
        protected void onCancelled() {
            try {
                socket.close();
            } catch (IOException e) {
                e.printStackTrace();
            }
        }
    }

    public void sendMessage(View view) {
        EditText editText = findViewById(R.id.
editText);
        TextView textView = findViewById(R.id.
textView2);
        String message = editText.getText().toString();
        textView.setText("Loading...");

        if (lo != null  lo.getStatus() == AsyncTask.
Status.RUNNING) {
            lo.cancel(true);
        }

        lo = new LongOperation();
        lo.execute(message);
    }
}
```

The server address and port are stored in the server_address and server_port variables. The “onCreate” method is called when the activity is created. It initializes the layout and retrieves the saved server address and port from the “SharedPreferences.” The shared preference MY_PREFS_NAME is used to store and retrieve the server address and port. If the saved values exist, they are displayed in the corresponding "TextView" elements. The initial values for the message and response are also set. The LongOperation class is an inner class that extends AsyncTask and handles the network operations.

In Android, long-running operations such as network requests should not be performed on the main user interface thread, as this can block the user interface and make the application unresponsive. AsyncTask is responsible for performing long-running operations in the background and updating the UI thread with the results. It establishes a socket connection with the server, sends the message, and reads the response.

The doInBackground method of LongOperation creates a socket and connects to the server address and port. The message passed as a parameter is sent to the server via the output stream of the socket. The response from the server is read from the input stream and returned as the result. The onPostExecute method updates the UI with the received response. The onPreExecute method is called before the background task starts and updates the server address and port from the UI elements. The onCancelled method is called when the task is canceled and closes the socket if it's still open. The sendMessage method is called when the Send button is clicked. It retrieves the message from the input field, sets the UI text to "Loading...," and starts the LongOperation AsyncTask with the message as a parameter. The UI elements are accessed using the findViewById method to update their contents.

The project utilizes the activity_main.xml file to create user interface layouts. The interface is straightforward, consisting of only three EditText boxes and one button as shown in Figure 10.1. Tapping the button sends a message from the input EditText (id: textView2) to the host (id: serverText) at a specified port (id: portText). To test the program, ensure that a simple echo TCP server is running with port 10000 as the listening port. When the button is tapped, the message is sent to the server through port 10000.

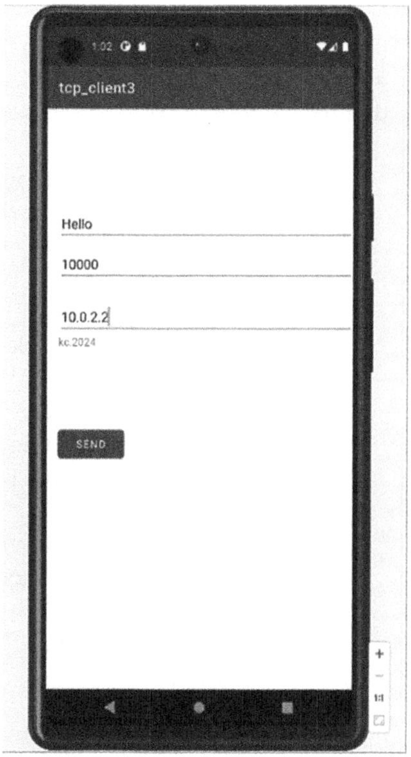

Figure 10.1 User interface of a TCP client

10.4.3 Python Code for Echo Server

The code shown in Code 10.2 is an implementation of an echo server in Python using socket programming to test with the TCP client program. The code starts by importing the `socket` library for socket programming and the `_thread` module for multithreading. To ensure thread-safe printing, a `print_lock`

Code 10.2 Echo Server Code Written in Python

Python code

```
# import socket programming library
import socket
from _thread import*
import threading

print_lock = threading.Lock()
# thread function
def threaded(c):
        while True:
                # data received from client
                data = c.recv(1024)
                if not data:
                        print('Welcome')

                        # lock released on exit
                        print_lock.release()
                        break
                # reverse the given string from client
                #data = data[::-1]

                # send back reversed string to client
                byt=data.decode()
                print(data)
                c.send(data)
        c.close()
```

```
def Main():
        host = ""
        # reserve a port on your computer
        # in our case it is 10000

        port = 10000
        s = socket.socket(socket.AF_INET, socket.
SOCK_STREAM)
        s.bind((host, port))
        print("socket binded to port", port)

        # put the socket into listening mode
        s.listen(5)
        print("socket is listening")

        # a forever loop until client wants to exit
        while True:

                # establish connection with client
                c, addr = s.accept()
                # lock acquired by client
                print_lock.acquire()
                print('Connected to :', addr[0], ':', addr[1])

              # Start a new thread and return its identifier
                start_new_thread(threaded, (c,))
        s.close()

if __name__ == '__main__':
        Main()
```

object from the `threading` module is created. Finally, the threaded function is defined. This function handles client connections by continuously receiving data from the client, reversing the received string, and sending it back to the client.

If no data is received, it prints "Welcome" and releases the print lock before breaking the loop and closing the connection. The `Main` function should be defined as follows: the `Main` function serves as the program's entry point. It sets the host (an empty string represents the local host) and reserves a port (in this case, port 10000). Next, it creates a socket object and binds it to the specified host and port. The server socket enters listening mode with a backlog of five connections.

The program enters a loop to accept incoming client connections. For each connection, it acquires the print lock, prints the client's address, and starts a new thread to handle the connection by calling the `` `threaded` `` function. Finally, the socket is closed. If the program exits the loop, it indicates that the server is no longer accepting connections, and the socket is closed to release resources. To execute the `` `Main` `` function, use the `` `if __name__ == '__main__':` `` condition. This ensures that the `` `Main` `` function is only executed when the script is run directly, not when it is imported as a module. Overall, the code sets up a server that listens for client connections and echoes back any data received from the client.

10.5 React Native

A native application refers to a software program designed specifically for a particular device platform and developed using a dedicated programming language. While native apps may incorporate web content, they are not intended to be web-based or employ HTML. Native apps are standalone programs that often make use of different functionalities on your device to operate. React Native [4] is developed by Facebook, which is a JavaScript-based cross-platform mobile application development framework. React Native uses React (React is a JavaScript library for building user interfaces) to write mobile applications for the iOS and Android platforms, it provides comparable performance and usability to that of native applications [5].

There are two ways to create a React Native app, by Expo CLI or React Native CLI. Expo CLI is a command-line interface tool for developers to create, develop, and manage cross-platform mobile apps using React Native. With Expo CLI, developers can easily build and test their apps on both iOS and Android devices without the need for platform-specific code or dependencies. It provides a streamlined development environment by abstracting away the complexities of setting up and configuring a React Native project. To view an Expo project, you must run the Expo CLI server for that specific project. The Expo CLI server acts as a local development server that hosts your app and provides a development environment for testing and previewing your application on various devices. This allows you to view your app in the Expo client app on your mobile device or a web browser using the Expo web interface. By running the Expo CLI server, developers can easily access and interact with their Expo project during the development process.

React Native CLI is a command-line interface tool used for creating and managing React Native projects. It enables developers to write platform-specific code and access native APIs directly. The React Native CLI is a built-in feature that helps users control and manage their projects locally. It provides a set of commands that allow users to initialize new projects, run them on simulators/emulators, and build the app binaries directly on their machines. This means they do not have to rely on a cloud service for building the app binaries, giving them more control and flexibility in the development process. On the other hand, the setup [6] is much more complicated. To build apps for Android, Android Studio needs to be installed, and many features need to be configured before getting started [7]. This process can be a bit complex, but the React Native CLI environment is better suited for professional app developers who need precise control over their React Native projects.

10.5.1 React Native Programming on MQTT Client

MQTT [8] is a communication protocol developed in 1999 by Dr. Andy Stanford-Clark of IBM and Dr. Arlen Nipper of Arcom. It provides a lightweight and reliable binary communication method, specifically designed for environments with narrow network bandwidth and limited power, such as those found in oil pipeline sensors and satellites. In November 2011, IBM and Eurotech donated the MQTT protocol to the Eclipse Foundation, which now manages the open-source project and is part of the Eclipse M2M Industry Working Group. By October 2014, MQTT had officially become an open OASIS international standard (Organization for the Advancement of Structured Information Standards), a nonprofit organization focused on developing e-commerce, network services, and electronic publishing.

Originally, MQTT stood for Message Queuing Telemetry Transport, but this abbreviation is no longer in use. The protocol's concise message format makes it ideal for IoT devices with limited processing resources and network capacity. Numerous MQTT libraries have since been developed for platforms such as Arduino (C/C++), JavaScript (Node.js, Espruino), and Python, along with an open-source MQTT server. This ecosystem facilitates the development of IoT applications and machine-to-machine (M2M) communication [9]. Notably, the instant messaging service of Facebook Messenger also utilizes the MQTT protocol.

The IoT powers many mobile applications, and React Native assists in developing numerous IoT applications. The MQTT protocol facilitates communication between IoT devices and supports bidirectional data exchange between clients and servers. MQTT communication operates as a publish-and-subscribe system, as shown in Figure 10.2. A device posts a message on a specific topic, and all devices subscribed to that topic will receive the message.

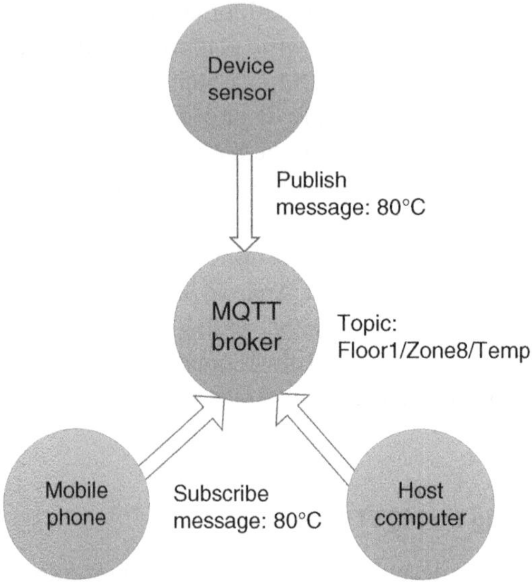

Figure 10.2 MQTT protocol

10.5.2 Explanation of the Code

The code provided in Code 10.3 is a React Native application written in JavaScript. It sets up an MQTT client using the `sp-react-native-mqtt` library and connects to an MQTT broker. The application also includes a simple user interface with a button and a text component.

Code 10.3 React Native Code for a MQTT Client

App.js code

```
import React from 'react';
import type {Node} from 'react';

import {
  SafeAreaView,
  ScrollView,
  StatusBar,
  StyleSheet,
```

```
  Text,
  useColorScheme,
  View,
  Button
} from 'react-native';

import {
  Colors,
  DebugInstructions,
  Header,
  LearnMoreLinks,
  ReloadInstructions,
} from 'react-native/Libraries/NewAppScreen';

import MQTT from 'sp-react-native-mqtt';

// Replace this URL with your own socket-io host, or
start the backend locally

export default function App() {

/* create mqtt client*/
    let mqttClient = null;
    MQTT.createClient({
      uri: 'mqtt://test.mosquitto.org:1883',
      //test.mosquitto.org:1883' is ok
      clientId: 'id08',
  })
    .then(function (client) {
      client.on('closed', function () {
        console.log('mqtt.event.closed');
      });

      client.on('error', function (msg) {
        console.log('mqtt.event.error', msg);
      });

      client.on('message', function (msg) {
        console.log('mqtt.event.message', msg);
      });
```

```
      client.on('connect', function () {
        console.log('connected');
        client.subscribe('Floor1/Zone8/Temp', 0);
        mqttClient = client;
      });

      client.connect();
    })
    .catch(function (err) {
      console.log(err);
    });

  return (
    <View style={styles.container}>
      <Button
       title="connect"
        color="#841584"
         onPress={() => {
           console.log(mqttClient);
            mqttClient.publish(' Floor1/Zone8/Temp ',
'80', 0, true);
          }}
/>

     <Text style={[styles.paragraph, { fontWeight:
"bold" }]}>
           @ 2023
      </Text>

    </View>
  );
}

const styles = StyleSheet.create({
  container: {
    flex: 1,
```

```
    backgroundColor: "#fff",
    alignItems: "center",
    justifyContent: "center",
  },
  paragraph: {
    fontSize: 16,
  },
  footnote: {
    fontSize: 14,
    fontStyle: "italic",
  },
});
export default MqttService.getInstance();
```

The `App` function is the entry point of the application and is exported as the default export. It returns JSX [10] (JavaScript XML) code that represents the user interface of the application. In the code, the MQTT client is created using the `MQTT.createClient` function from the `sp-react-native-mqtt` library. The `uri` property specifies the URL of the MQTT broker (i.e., test.mosquitto.org [11]), and the `clientId` property sets the client ID for identification. The client then sets event listeners for the `closed`, `error`, `message`, and `connect` events. Upon successful connection, the client subscribes to the `Floor1/Zone8/Temp` topic and assigns the MQTT client object to the `mqttClient` variable.

The user interface is defined inside a `View` component with the `styles.container` style. It contains a `Button` component that triggers a function when pressed. In this case, the function logs the `mqttClient` object and publishes a message to the `Floor1/Zone8/Temp` topic with a payload of "80." The `styles` object defines the styles for the application using the `StyleSheet.create` function, including styles for the container, paragraph, and footnote.

The code also includes a module that exports a static method, `getInstance`, of the `MqttService` class, which implements the singleton pattern. Whenever the `MqttService` is imported, it either returns the existing instance or creates a new one, even if one does not already exist. Overall, this code sets up an MQTT

Figure 10.3 User interface of a MQTT client

client, connects to an MQTT broker, and provides a button to publish a message on a specific topic. The user interface displays a text component and a button.

The program features a very simple user interface with only one button, as shown in Figure 10.3. When the user taps the button on the phone, the message "80" will be published to the broker's topic `` `Floor1/Zone8/Temp` ``. Within the project folder, if the command `` `react-native run-android` `` is executed, a startup screen will appear, as shown in Figure 10.4.

The message has been published and can be viewed using MQTT Explorer, as shown in Figure 10.5.

10.5.3 Advantages and Disadvantages of React Native

One advantage of using React Native is that those familiar with React can get started quickly. Additionally, the use of the JavaScript library can be both convenient and powerful. Furthermore, if Node.js is used as the backend, both the front and back ends can uniformly utilize TypeScript [12]. However, one potential issue with React Native is that some libraries may not be as well-maintained as desired.

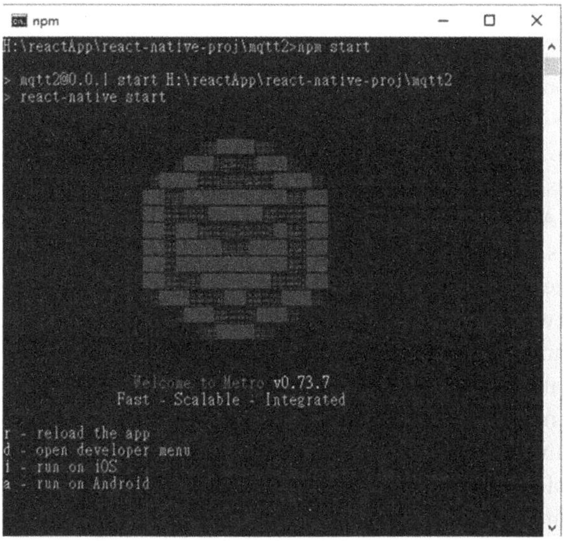

Figure 10.4 Startup screen

Figure 10.5 Result from MQTT Explorer

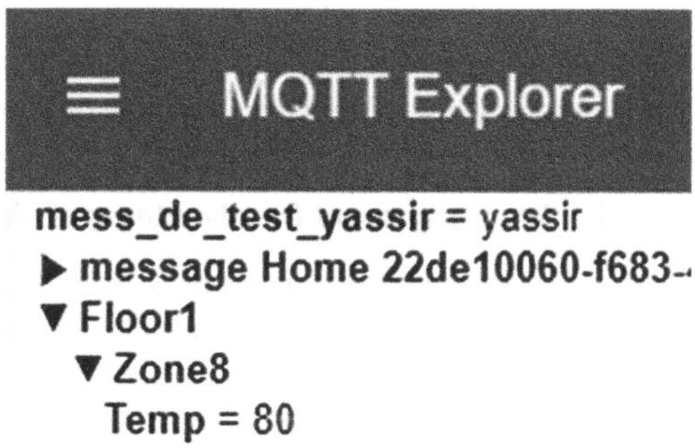

Bugs can arise later, either from updates to iOS or Android versions or from new updates to React Native, even if the library was reliable when initially released. In such cases, users may need to address the issues themselves.

Another potential drawback is the reliance on third-party libraries, which can be extensive. While React Native is a suitable platform for building simple applications, it may not be the ideal choice if complex functionality and high performance are priorities. Users should be aware that adding more functionality may impact performance.

10.6 Flutter

Flutter [13] is an open-source cross-platform application SDKs developed by Google for Android, iOS, Windows, macOS, and Linux desktops. It is a powerful framework [14] that enables developers to create Android and iOS apps using a single codebase. Flutter consists of three main layers: the top layer provides a reactive user interface for both Android and iOS platforms, the second layer includes Flutter's native components, and the third layer is responsible for rendering the user interface code into pixels.

Unlike many other frameworks, Flutter is written entirely in the Dart programming language [15]. While most programming engines use languages like C++ or Java for Android-specific features and Objective-C for iOS-specific features, Flutter uses Dart to avoid performance issues associated with compiled programming languages. This approach eliminates the need for a JavaScript bridge. Dart Native is an "Ahead of Time" compiler that produces native code for multiple platforms, allowing Flutter to communicate directly with the platform without relying on a JavaScript bridge for content conversion. Additionally, Flutter compiles native code, resulting in faster app startup times.

In Flutter, widgets [16] play a central role. Widgets are elements that control and impact the app's view and interface, similar to React components. They are essential building blocks for creating UI components, making it easy to develop and customize the user interface with just a few lines of code.

10.6.1 Advantages and Disadvantages of Flutter and React Native

Flutter's development environment is generally easier to set up compared to that of React Native. Additionally, Flutter's hot reload feature is more responsive than React Native's fast refresh. Flutter provides excellent component debugging, analysis, and inspection tools, and its built-in end-to-end testing function is more capable than React Native's Detox. Furthermore, Flutter's command-line interface (CLI) is among the best in the industry, particularly the Flutter Doctor command, which allows developers to manage their simulators and emulators directly through the CLI.

Additionally, Flutter's upgrade experience is superior. Developers can run `flutter create` directly in the existing application, and it will rebuild everything according to the new version. In contrast, many of React Native's tools are not as elegant or complete as Flutter's. The most challenging aspect remains the upgrade experience, and developers who have used React Native in recent years are likely to have faced similar issues. However, the situation is gradually improving. Several industry leaders from Microsoft have discussed ideas for enhancing React Native's tools and developer experience in various interviews.

Furthermore, Expo significantly improves the developer experience compared to React Native. With the Expo service, developers not only get all the functionality of React Native but also a better upgrade experience and integration tooling. If there are no plans to use custom native code, Expo Go is a great way to quickly share a build with others without compiling it. Regarding performance, it is challenging to make a direct comparison between software frameworks, but both Flutter and React Native are generally "fast enough" for most use cases.

10.7 Swift

Swift [17] is a programming language developed by Apple for building applications on iOS and macOS platforms. Since its introduction in 2014, it has gained significant popularity among developers. One of the main reasons for its widespread adoption is that Swift simplifies the complex programming structures found in Objective-C and is designed to be accessible to individuals without a programming background.

By incorporating the best features of Objective-C, Python, Ruby, and JavaScript, Swift enables developers to write simpler and more intuitive code that achieves the same functionality as more complex code in other languages. This makes development more efficient and helps streamline the process of implementing desired features.

10.7.1 Xcode

Xcode is an IDE developed by Apple for creating software for macOS, iOS, watchOS, and tvOS. It offers a complete set of tools, including a code editor, debugger, interface builder, and simulator, to simplify the development process. Xcode supports various programming languages, with Swift as the primary language for developing iOS and macOS applications. It provides frameworks and libraries to assist developers in creating interactive and feature-rich applications.

10.7.2 Development Environment

If a Mac computer is not possessed, limitations may be faced in using Swift to develop iOS applications. Additionally, it is worth noting that working with the iPhone can sometimes be challenging, as Xcode needs to be paired with the corresponding phone version, and compatibility issues may arise with very new or old devices. However, it is worth mentioning that efforts have been made by the Swift Project to introduce new downloadable Swift toolchain images for Windows, which can be helpful for learning purposes. It is important to note that while these tools provide opportunities to explore Swift on non-Mac platforms, they may have limitations when it comes to full-fledged iOS app development.

10.7.3 Xcode Programming on Simple TCP Client

A TCP client program utilizing the native Swift package is used for demonstration. Various ways to establish a socket connection are available, but for this demonstration, the default iOS frameworks—such as Network, Input, and Output Stream—are employed. First, Xcode is opened, and a welcome dialog is displayed. From here, "Create a new Xcode project" is selected to start a new project. A variety of project templates are shown for selection. An "App" application is created by selecting iOS at the top of the dialog.

In the Xcode project, the file `ViewController.swift` is opened. Near the top, the following line needs to be added:

```
import UIKit
import Network
var connection: NWConnection!
```

The complete Swift Xcode code for a TCP client implementation using the Network framework in iOS is shown in Code 10.4. This implementation connects to a TCP server at a given IP address and port, sends a message, and receives data from the server.

Code 10.4 Xcode Code for a TCP Client

```
//  ViewController.swift
//  tcpClient
//  Copyright © 2023 apple. All rights reserved.

import UIKit
import Network

class ViewController: UIViewController {
    var connection: NWConnection?
    let tcpPort: NWEndpoint.Port = 10000
    let tcpServerAddress: NWEndpoint.Host = "XXX.XXX.
XXX.XXX"
    let queue = DispatchQueue(label: "TCP Client Queue")

    override func viewDidLoad() {
```

```
        super.viewDidLoad()
       // Do any additional setup after loading the view.
    }

    @IBAction func connection(_ sender: Any) {
        connectToTCPServer()
    }

    func connectToTCPServer() {
        let tcpOptions = NWProtocolTCP.Options()
        tcpOptions.noDelay = true

        let parameters = NWParameters(tls: nil, tcp:
tcpOptions)
        connection = NWConnection(to: NWEndpoint.hostPort
(host: tcpServerAddress, port: tcpPort), using:
parameters)

        connection?.stateUpdateHandler = { [weak self]
newState in
            switch newState {
            case .ready:
                print("Socket State: Ready")
                self?.sendMessage()
                self?.receiveData()
            default:
                print("Socket State: \(newState)")
            }
        }

        connection?.start(queue: queue)
    }

    func sendMessage() {
        print("Sending data...")
        let message = " hello world "
        guard let data = message.data(using: .utf8) else
{ return }
```

```
        connection?.send(content: data, completion:
.contentProcessed({ error in
            if let error = error {
                print("Error sending data: \(error)")
            } else {
                print("Data sent successfully")
            }
        }))
    }

    func receiveData() {
        connection?.receive(minimumIncompleteLength: 1,
maximumLength: 65536) { [weak self] (data, _,
isComplete, error) in
            if let data = data, !data.isEmpty {
                print("Received data: \(data.count)
bytes")
                // Process the received data here
            }

            if isComplete {
                print("Receive is complete")
            } else if let error = error {
                print("Error receiving data: \(error)")
            } else {
                self?.receiveData()
            }
        }
    }
}
```

The code defines a `ViewController` class that serves as the main entry point of the application. The `ViewController` class has several properties, including an optional `NWConnection` instance to represent the TCP connection, the port number and IP address of the TCP server, and a `DispatchQueue` to manage the TCP client operations. The `viewDidLoad()` method is called when the view controller is loaded, and it performs any additional setup required for the view.

The `connection(_:)` method is an IBAction that is called when a button or other UI element is triggered. This method calls the `connectToTCPServer()` function to establish the TCP connection. The `connectToTCPServer()` function sets up the TCP connection using the `NWConnection` class. It configures the TCP options, creates an `NWConnection` instance, and sets up the state update handler. When the connection is ready, it calls the `sendMessage()` and `receiveData()` functions.

The `sendMessage()` function sends a message (in this case, "hello world") to the TCP server. It converts the message to `Data` and sends it using the `send(content:completion:)` method of the `NWConnection` instance. The `receiveData()` function receives data from the TCP server. It uses the `receive(minimumIncompleteLength:maximumLength:completion:)` method of the `NWConnection` instance to receive data. If data is received, it is printed. If the reception is complete or an error occurs, it handles the respective cases.

Overall, this code sets up a TCP client connection, sends a message to the server, and receives data from the server. It demonstrates the use of the Network framework in iOS to handle TCP communication. The user interface, as shown in Figure 10.6, is simple, featuring a single button to connect and send a message to a host. The response from the Echo server is shown in Figure 10.7.

Figure 10.6 User interface of the TCP program

10.8 Other Development Considerations

As a summary of the steps illustrated through the examples, smartphone application development encompasses

```
Python 3.8.2 (tags/v3.8.2:7b3ab59, Feb 25 2020, 23:03:10) [MSC v.1916 64 bit (AM
D64)] on win32
Type "help", "copyright", "credits" or "license()" for more information.
>>>
= RESTART: C:\Users\Rider\Desktop\tmp_book_resource\Py_server\mt_server_3a_echo.
py
socket binded to port 10000
socket is listening
Connected to : 219.77.204.152 : 49261
b'hello world'
|
```

Figure 10.7 Response from the echo server

the full life cycle of creating software applications designed to run on mobile devices such as smartphones and tablets. The process begins with the planning and design stage, where the app's purpose, target audience, and key features are defined. This is followed by designing the user interface and user experience, ensuring an intuitive and engaging interaction for the end users. Determining the appropriate mobile platforms to develop for, such as iOS, Android, or cross-platform options, is an important decision at this stage, as is selecting the programming languages, frameworks, and development tools that will be utilized.

The coding and development phase involves implementing the app's functionality using languages like Swift, Objective-C, Java, or Kotlin. Developers leverage mobile-specific APIs and frameworks, such as the iOS frameworks or Android SDK, to integrate the app with the device's hardware and sensors, including features like the camera, GPS, and accelerometer. Ensuring the app is responsive and optimized for different screen sizes and resolutions is also a crucial consideration during this stage.

The event-driven nature of Android Studio programming aligns well with the event-based nature of mobile applications, where user interactions and external events drive the application's behavior. This architectural approach helps ensure a responsive and smooth user experience. In contrast, the sequential nature of C, as discussed in Chapter 7, is better suited for applications that have a more predictable flow of control, such as system-level software, embedded systems, or performance-critical algorithms. Both event-driven and sequential programming have their strengths and are suited for different types of applications and problem domains. The choice between them depends on the specific requirements of the project and the developer's familiarity with the different programming paradigms. The same principles regarding event-driven and sequential programming apply to Xcode development for iOS and macOS applications, just as they do for Android Studio and the Android platform.

Once the app is built, thorough testing for functionality, performance, and user experience is conducted. The completed app is then submitted for review and approval to the relevant app stores, such as the Apple App Store or Google Play, before being deployed for users to download and utilize. Ongoing maintenance and updates are essential, as developers release bug fixes, improve functionality, and adapt the app to platform changes based on user feedback and market trends.

Throughout the entire smartphone application development process, close attention is paid to creating a seamless, intuitive, and engaging user experience that meets the needs and expectations of the target audience. The iterative nature of this process, from planning and design to coding, testing, deployment, and maintenance, ensures that the app remains relevant, competitive, and responsive to the evolving mobile landscape. Successful smartphone app development requires a combination of technical skills, design expertise, and a deep understanding of the target mobile platforms and user needs. Developers often work in cross-functional teams, collaborating with designers, product managers, and quality assurance personnel to deliver high-quality, user-friendly mobile applications.

References

1 Sills, B. et al. (2022). *Android Programming: The Big Nerd Ranch Guide (Big Nerd Ranch Guides).* Addison-Wesley Professional.

2 "What is Gradle? Why do we use Gradle? Explained," last modified February 5, 2024, https://www.simplilearn.com/tutorials/gradle-tutorial/what-is-gradle

3 "What is Kotlin?" last modified February 5, 2024, https://www.techtarget.com/whatis/definition/Kotlin

4 Eisenman, B. (2017). *Learning React Native: Building Native Mobile Apps with Javascript.* O'Reilly Media.

5 "What is a native app?" last modified February 5, 2024, https://uxcam.com/glossary/native-app/

6 "Setting up the development environment," last modified February 5, 2024, https://reactnative.dev/docs/environment-setup

7 "Get started developing for Android using React Native," last modified February 5, 2024, https://learn.microsoft.com/en-us/windows/dev-environment/javascript/react-native-for-android

8 Hillar, G.C. (2017). *MQTT Essentials - A Lightweight IoT Protocol: Send and Receive Messages with the MQTT Protocol for Your IoT Solutions.* Packt Publishing.

9 "Machine-to-Machine (M2M)," last modified February 5, 2024, https://www.techtarget.com/iotagenda/definition/machine-to-machine-M2M

10 "ReactJS introduction to JSX," last modified February 5, 2024, https://www.geeksforgeeks.org/reactjs-introduction-jsx/
11 "test.mosquitto.org," last modified February 5, 2024, https://test.mosquitto.org/
12 "TypeScript introduction," last modified February 5, 2024, https://www.w3schools.com/typescript/typescript_intro.php
13 Windmill, E. (2020). *Flutter in Action*. Manning Publications.
14 "Flutter architecture," last modified February 5, 2024, https://www.javatpoint.com/flutter-architecture
15 "Dart overview," last modified February 5, 2024, https://dart.dev/overview
16 "Introduction to Widgets," last modified February 5, 2024, https://docs.flutter.dev/development/ui/widgets-intro
17 Sahar, A. (2023). *iOS 17 Programming for Beginners - Eighth Edition: Unlock the World of iOS Development with Swift 5.9, Xcode 15, and iOS 17 - Your Path to App Store Success*. Packt Publishing.

Part IV

Putting into Practice

11

Practical Design Examples in Modern Wireless System Design

11.1 Introduction

In this chapter, practical design examples will be presented to illustrate the importance of the design concepts and techniques discussed in the previous chapters regarding Internet of Things (IoT) and wireless solution development. These concepts are essential for understanding how to effectively implement technology in real-world applications. Two projects related to IoT and wireless technologies for smart cities will be discussed in detail. As urban areas continue to grow, smart cities require innovative technologies to enhance efficiency, sustainability, and economic growth. These technologies are designed to optimize various systems, including energy consumption and transportation, while also promoting the use of renewable energy and sustainable practices.

IoT plays a crucial role in this context, enabling the monitoring and management of critical infrastructure. By connecting devices and systems, IoT facilitates real-time data collection and analysis, leading to more informed decision-making and improved service delivery. The two specific projects explored in this chapter are the Smart Lamp Post (SLP) and the Smart Radio Frequency Identification System (SRFID). The SLP integrates lighting management with environmental monitoring, while the SRFID System enhances asset tracking management. These examples demonstrate the seamless integration of hardware, firmware, webpage design, artificial intelligence (AI), and software development to create innovative solutions tailored for urban environments within smart cities.

Through these case studies, readers will gain insights into how the suggested concepts and techniques can be applied in their own projects. This knowledge will empower them to develop effective IoT and wireless solutions that address the unique challenges faced by modern urban areas.

Introduction to Wireless System Design: From Circuits to Web-based Applications, First Edition.
Henry Lau, Ludy Liu, and Keith Chan.

11.2 IoT Technologies for Smart Cities

Smart cities require new technologies for several reasons [1]. Firstly, these technologies are crucial for enhancing efficiency within the city. By optimizing various systems such as energy consumption, transportation, waste management, and infrastructure usage, new technologies enable cities to operate more effectively. For example, sensors and data analytics can be utilized to monitor and manage energy usage, reducing waste, and improving resource allocation. Secondly, new technologies play a vital role in promoting sustainability in smart cities. Aiming to reduce their ecological footprint, smart cities employ innovative technologies to achieve environmental goals. This includes implementing renewable energy generation, smart grid systems, and optimizing transportation networks to minimize carbon emissions. Additionally, these technologies facilitate sustainable practices in areas like water management and waste reduction, contributing to a greener and more environmentally conscious city.

Economically, smart cities that adopt new technologies attract businesses and investments. The innovative infrastructure and digital capabilities of these cities appeal to startups and established businesses, driving economic growth and creating job opportunities. Furthermore, the improved quality of life resulting from new technologies makes the city more attractive to both residents and businesses, fostering a thriving economic ecosystem.

IoT technology [2, 3] plays a significant role in the development of smart cities. It enables the interconnection and communication between various devices and systems, allowing cities to optimize resource utilization, enhance services, and improve the overall quality of life for their residents. One area where IoT is applied in smart cities is smart infrastructure. By deploying IoT sensors throughout the city, critical infrastructure such as roads, bridges, buildings, and utility systems can be monitored and managed more effectively. These sensors can detect issues like structural damage, leaks, or faulty equipment, enabling proactive maintenance and reducing the risk of failures or accidents.

In order to support the Smart City initiatives in Hong Kong and promote sustainability while creating a smarter airport, two IoT projects were developed: the SLP and the SRFID System. This chapter will explore the design specifics of these projects in depth. The SLP project utilized the latest IoT technology to perform remote monitoring and management of SLPs. Centralized control systems allowed administrators to monitor the status, performance, and energy consumption of individual lamp posts. They could also remotely configure settings, schedule maintenance, and detect faults for efficient operations. The SRFID System project aimed to enhance efficiency and security within the airport environment. RFID technology [4] was utilized to track and manage the movement of people and assets throughout the airport premises. RFID tags or cards were assigned to individuals and objects, such as baggage, vehicles, and airport staff. Passengers benefited from the SRFID System through streamlined check-in processes and automated baggage handling. RFID tags on luggage enabled real-time tracking and monitoring, reducing the risk of lost or mishandled baggage. Additionally, RFID-enabled access control systems enhanced security by providing secure and efficient identification of individuals entering restricted areas.

11.3 Smart Lamp Post

Traditionally, street lamp posts have relied on AC power for their energy source. However, there is a growing shift toward solar-powered LED lamp posts in order to promote energy efficiency and environmental protection. These new lamp posts utilize LED lighting technology, which is more energy-efficient than traditional lighting sources, and are powered by solar panels. While these solar-powered LED lamp posts offer benefits such as reduced energy consumption, they also introduce additional components and devices that may experience failures over time. To address this issue, an Automatic Monitoring System (AMS) was implemented to oversee the operation of the lamp posts and to promptly notify the system administrator in case of any faults or issues. The SLP project, hosted by the Electrical and Mechanical Services Department (EMSD) of Hong Kong and implemented by City University of Hong Kong and Lexiwave Technology (Hong Kong) Limited, served as a pilot study program. This project leveraged IoT technology to enable automatic monitoring of the solar-driven LED lamp posts.

The project involved designing and building a sensing unit, which was then installed in 23 selected SLPs designated for the pilot program in a rural area. The original design, shown in Figure 11.1, has been modified for illustrative purposes in this book. The sensing unit consisted of a control module, a temperature sensing module, and a battery diagnostics module.

The sensing unit monitored the operation of the lamp post and measured all critical operational parameters, such as lamp current and battery charging current. The sampled data was broadcast to a local LoRa Gateway [5] through LoRa radio, as shown in Figure 11.2. The collected data from all the lamp posts was then relayed to a computer server through the 4G cellular network by the gateway in batches for storage and analysis. This approach was more cost-effective than sending data from each lamp post individually.

Users could monitor this data and basic analytical results through a dashboard in a web portal via a PC or mobile device. The database could be exported as Comma-Separated Values or Excel files for further analysis.

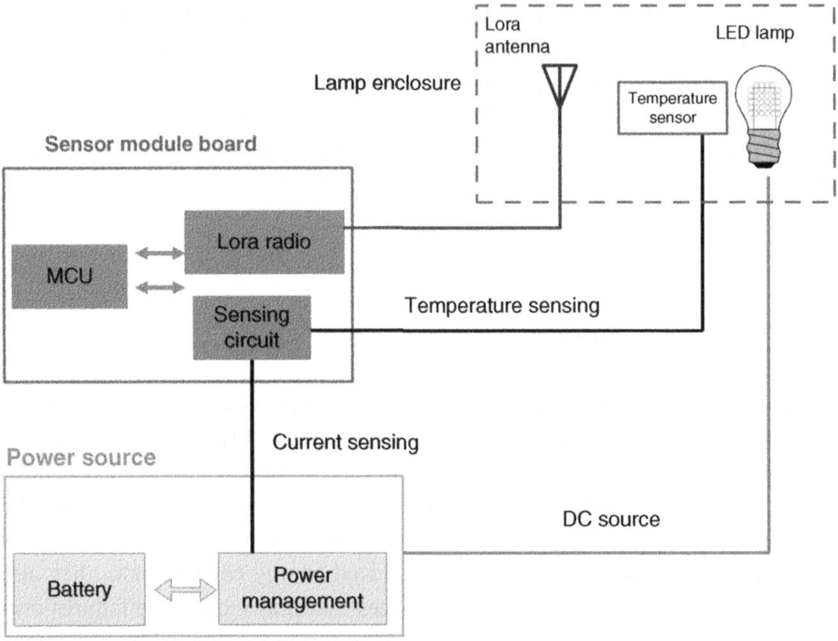

Figure 11.1 Sensing device of the SLP system

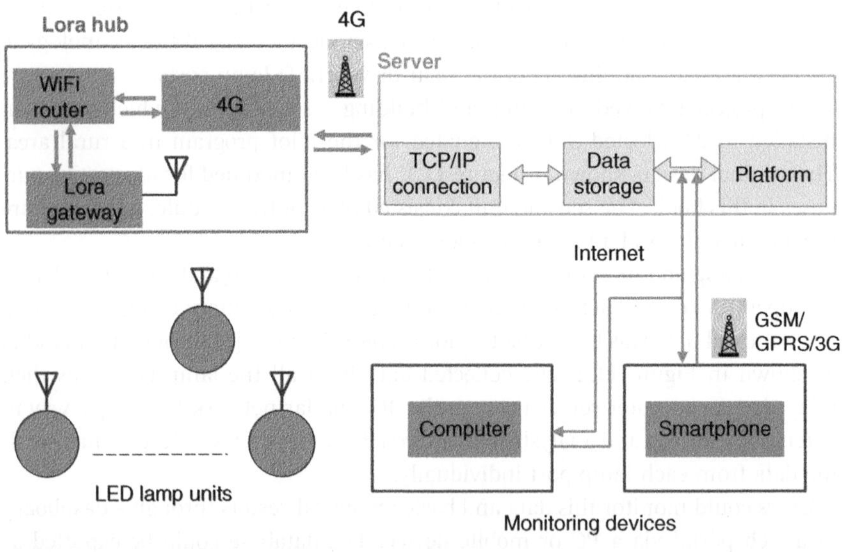

Figure 11.2 Complete SLP system

11.3.1 Hardware Design

Figure 11.1 shows the sensing unit, which consists of a control module, a temperature sensing module, and a battery diagnostics module. The battery diagnostics module was responsible for monitoring the DC current supplied to the LED lamp, the temperature of the power chassis of the lamp post, and the battery life of the internal power source. This information helped users detect any malfunctioning lamp posts by utilizing the current sensing module. Additionally, the battery diagnostics module and the temperature sensing module contributed to predicting the battery's operational status.

Traditional technology typically only allows for the measurement of the instantaneous power output of a battery, making it challenging to accurately predict its lifetime. However, the battery diagnostics module incorporated a special algorithm that enables a more precise prediction of the battery's working condition by considering its state of charge (SOC) and state of health (SOH) [6, 7]. Both the battery diagnostics and control modules used a microprocessor to perform diagnostic and control functions, respectively. A microprocessor (MCU) was preferred over a single-board computer due to its low power consumption and low cost, as discussed in Chapter 7.

Both the battery diagnostics module and the control module used an MCU to perform their respective functions. The chosen MCU was the STM32 series [8] from ST Microelectronics. It was a high-density performance line Arm-based 32-bit MCU with 256–512 KB of Flash memory and multiple communication interfaces, such as Serial Peripheral Interface, Universal Asynchronous Receiver-Transmitter (UART), and Controller Area Network. Its computational power was more than sufficient for this type of IoT application, which involved very focused tasks. It was preferred over other single-board computers, such as the Raspberry Pi [9] or the Beagle Board [10], due to its low power consumption, as the entire monitoring unit was powered by a solar panel. The typical current used was only a few milliamperes for STM32 MCUs.

The control module collected the measuring parameters from the diagnostics module via the UART communication interface. The data was then sent to a LoRa gateway through LoRa [5] radio communication. LoRa communication was chosen due to its long-range capability, as it uses Chirp Spread Spectrum (CSS) modulation for superior receiving sensitivity (see Chapter 1). The 23 posts were spread over a distance of 7 km, as shown in Figure 11.3, which could be served by only one gateway. A LoRa wireless communication module [11] from AcSIP Technology Corporation was used for the LoRa transmission.

Using wireless communication modules is very common in developing IoT systems. There are several reasons why it is preferable to use these modules instead of building one's own. One significant factor is time and cost efficiency. Developing a

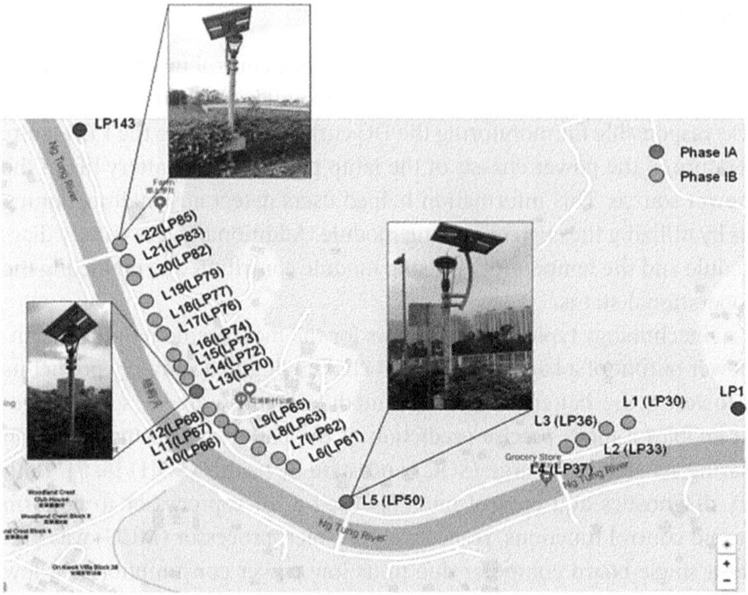

Figure 11.3 Location of the Smart Lamp Posts (SLP)

wireless communication system from scratch requires substantial time, expertise, and resources. It involves designing and prototyping the necessary hardware, developing the firmware and software, and conducting extensive testing and certification processes. By utilizing pre-built wireless communication modules, one can save a considerable amount of time and significantly reduce development costs.

Another crucial aspect is the reliability and performance of wireless communication modules. These modules are designed and manufactured by specialized companies with expertise in wireless technologies. They undergo rigorous testing and certification processes to ensure their reliability, performance, and compliance with industry standards. By using established modules, one can benefit from the expertise and track record of these companies, resulting in a more reliable, and robust solution.

The controller PCB incorporated the MCU, LoRa module, and their associated components. Despite the relatively low component count, the PCB design adhered to proper techniques and guidelines (previously discussed in Chapter 6) to ensure the creation of a functional RF board. To accommodate the small number of components, a two-layer PCB design was deemed adequate, as depicted in Figure 11.4. The component placement was strategically determined with RF considerations in

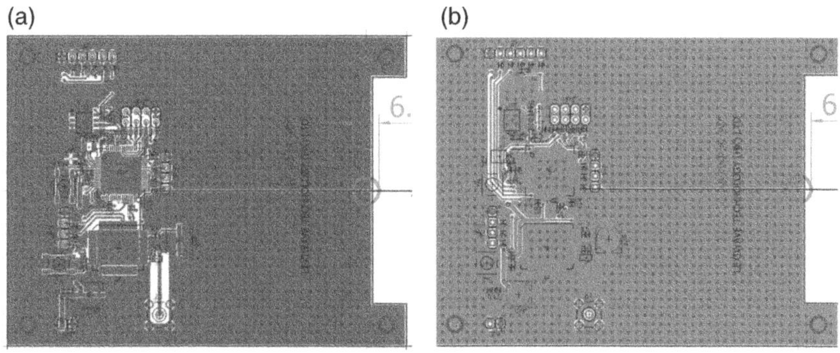

Figure 11.4 PCB layout of the controller board (a) top layer, and (b) bottom layer

mind. The presence of a comprehensive ground plane, supported by a sufficient number of ground via holes, further enhances the board's performance.

Next, the RF output of the LoRa module on the controller board was connected to a monopole antenna located atop the solar lamp (as illustrated in Figure 11.5) using an RF coaxial cable. The selection of a monopole antenna was based on its high efficiency and relatively compact size, taking into account the presence of a ground plane (as explained in Chapter 5). Since the casing of the solar lamp was constructed from metal, it served as the ground plane for the monopole antenna.

Figure 11.5 Monopole antenna on a metal base

The LoRa gateway was installed in a central position among all the lamp posts and utilized the same monopole antenna. Subsequently, the gathered data from all the lamp posts was transmitted to the server computer via a 4G network using a 4G modem.

11.3.2 Firmware Design

The embedded software used for development was the C programming language using STM32CubeIDE [12] which is an all-in-one multi-OS development tool being part of the STM32Cube software ecosystem. As mentioned in Chapters 7 and 8, C or C++ programming is used for embedded systems due to its hardware direct dependence for real time and fast response.

In this application, sequential programming was chosen over the Real-Time Operating System (RTOS) approach due to the low number of tasks involved. The MCU only needed to establish communication with the battery diagnostic module and the LoRa module. While this approach offered simplicity, it was still recommended to generate status logs from the MCU for debugging purposes. This can be achieved by printing the logs through the UART port, if available. In cases where all the UART ports are occupied, alternative options can be explored, such as obtaining status logs from the Joint Test Action Group (JTAG) port, which is used for programming and debugging.

11.3.3 Webserver Design

As introduced in Chapter 9, a web server is a software application or computer system responsible for delivering web content over the Internet or an Intranet. It serves as the foundation for hosting websites, web applications, and other online services. When a user requests a webpage or resource through a web browser, the web server processes the request and sends the appropriate response back to the user's browser. The computer system can be a cloud server or a physical server. In this project, a physical computer was used for the server since cloud server services were still expensive when the system was developed. The server was equipped with the XAMPP web server package, which provides hosting for webpages using Apache, database management with MySQL, and PHP for server-side web programming.

Additionally, a Python program was installed on the server to handle the Message Queuing Telemetry Transport (MQTT) [13] data messages sent by the LoRa gateway. In addition to the Python script, an MQTT program called Mosquitto was also installed to act as the MQTT broker. Figure 11.6 illustrates the MQTT data flow from the SLP, where the LoRa gateway served as the publisher, Mosquitto acted as the MQTT broker, and the Python program functioned as the subscriber. Furthermore, the Python program was responsible for posting the final data into the database.

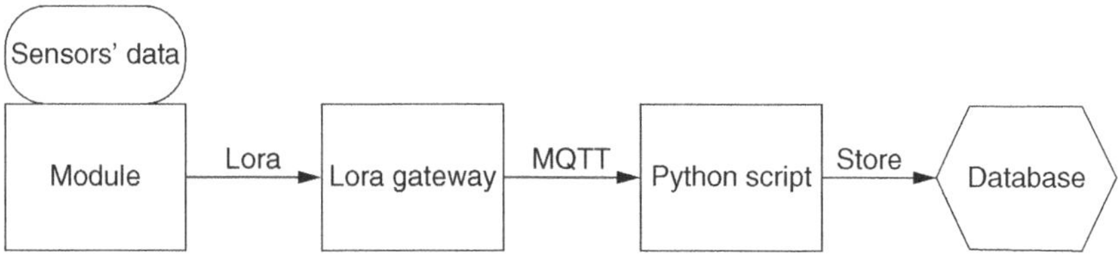

Figure 11.6 Brief block diagram for data flow

11.3.4 Webpage Design

The development of a webpage involves a series of steps that can be categorized into various phases. Initially, the purpose and scope of the webpage are defined. This includes identifying the target audience and clarifying the goals, content, and functionality requirements. Factors such as design style, interactivity, and the utilization of specific features or technologies are also taken into account. Once these considerations are made, an outline is created to structure the main sections and pages, as depicted in Figure 11.7.

The first page of the website was the Login page, which allowed users or administrators to access the system. An essential component of a web portal for an IoT system was the dashboard, which provided users with charts and graphs to monitor and control IoT devices. When designing the dashboard, it was important to

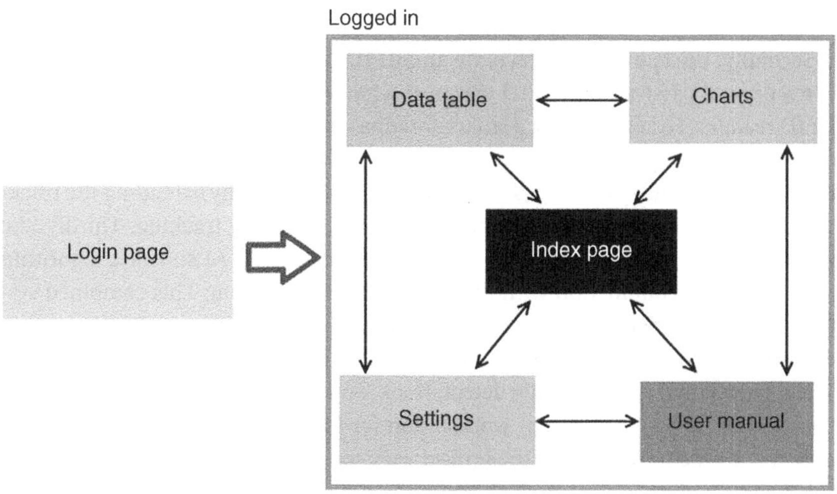

Figure 11.7 Outline of the webpages

consider aspects such as real-time data visualization, device management, analytics, and user interaction.

Bootstrap is a Hypertext Markup Language (HTML), CSS, and JavaScript front-end framework used for implementing the design of webpages. Its primary objective is to enable websites to automatically adjust their layout to different screen sizes, ensuring responsiveness. While Bootstrap provides a convenient solution, additional programming with HTML, JavaScript, Hypertext Preprocessor (PHP), and CSS is also employed for further customization. These languages are particularly useful for tasks such as data exchange between pages and accessing the database. Figure 11.8 showcases some of the developed webpages. In IoT dashboards, charts and tables are indispensable elements for effective data visualization and presentation.

11.4 Smart RFID System

This collaborative project took place in Hong Kong, involving the EMSD of Hong Kong and Lexiwave Technology (Hong Kong) Limited. It was funded by EMSD and aimed to develop an SRFID Antenna System for indoor passive Ultra High Frequency (UHF) RFID tag tracking. The project had three primary objectives. Firstly, it focused on developing new RFID tags that were compatible with the SRFID Antenna System. These specialized tags not only improved tracking efficiency but also expanded the coverage of the RFID system. By incorporating these tags, the system became more versatile and adaptable to various tracking scenarios.

Secondly, the team aimed to create an SRFID Antenna System that could track conventional passive UHF RFID tags over a longer range compared to traditional RFID readers. To achieve this, a smart antenna with beamforming technology was developed. This system could automatically detect, track, and communicate the tag's identity to the system workstation for further actions. By extending the tracking range, the system enhanced the efficiency of RFID tag tracking. Thirdly, the team established a Convolutional Neural Network (CNN) AI software algorithm that worked in conjunction with the SRFID Antenna System. This combined system had an extended coverage range compared to conventional RFID readers, enabling the tracking of conventional passive RFID tags in indoor environments. The system could automatically detect, track, and trigger other systems to provide the tag's identity to the system workstation for further actions. It is important to note that the actual designs of the project were modified for illustrative purposes in this book.

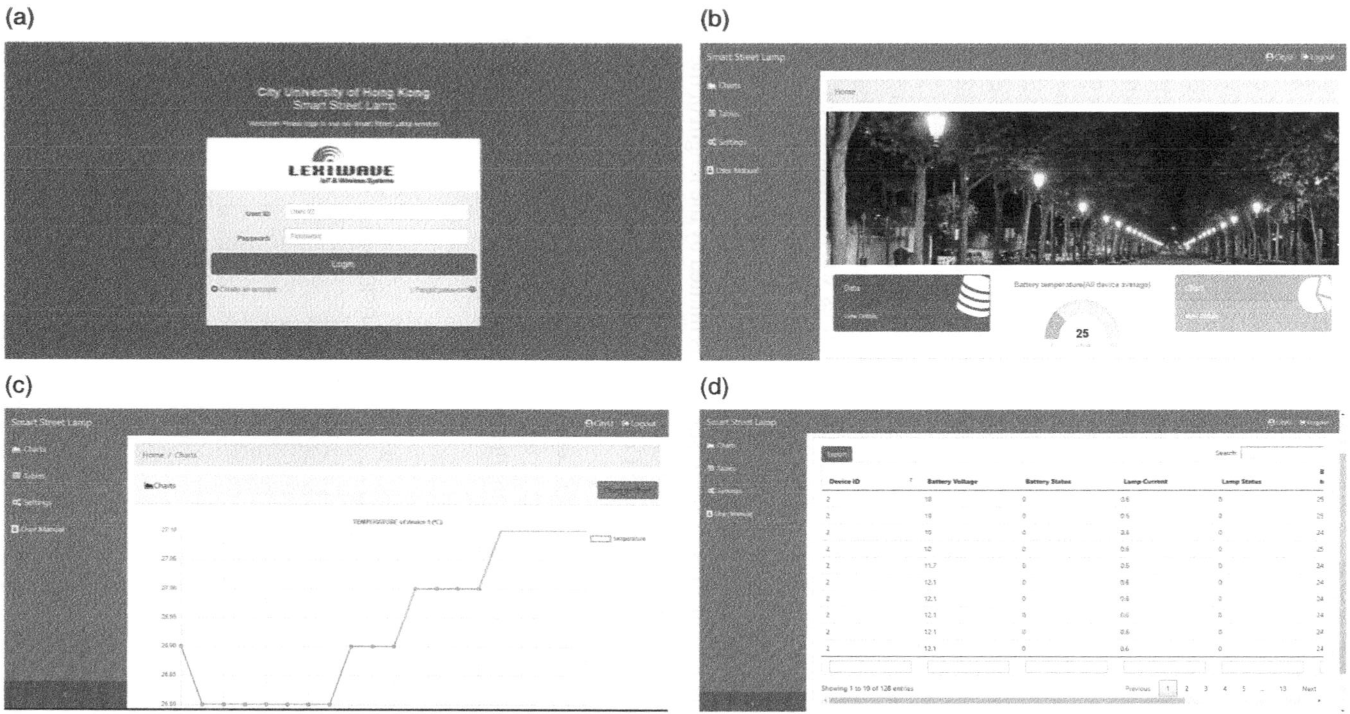

Figure 11.8 Developed webpages (a) login, (b) main index, (c) chart, and (d) table

11.4.1 Hardware Design

11.4.1.1 Passive RFID Tag Design

One of the main objectives of the project was to improve the coverage of conventional passive RFID tags. Much effort was spent on enhancing read sensitivity and conducting tests. Since the detection range was the primary indicator of read sensitivity, a comparison of existing passive RFID tags of different sizes and shapes on the market was performed using a standard RF reader (single channel) with a non-beamforming antenna for read distance measurement in benchmark sensitivity tests. The block diagram of the test setup is shown in Figure 11.9, and the equipment used and the testing location are shown in Figure 11.10. The testing location was along Corridor A outside an office, with a length of 30 m. A single-channel RFID reader with a standard single-patch RFID antenna was used to check the maximum detection distance.

The first passive RFID tag prototype used NXP Semiconductors' SL3S4011 RFID chip [14]. The antenna designed was a traditional differential inverted-F type [15]. Three tags of different sizes and antenna structures were fabricated for performance comparison. The designed RFID tags namely, SL-A, SL-B, and SL-C along with some other tags on the market to be tested and compared, are shown in Figure 11.11. It was found that the maximum detection distance for the tags on the market was around 30 m, as this distance was limited by the length of the corridor, while the first prototype achieved a distance of about 24 m. The distance test was repeated in Corridor B (next to Corridor A), which has a length of 40 m, as shown in Figure 11.12.

Based on the test results in both Corridors A and B, it was suspected that the RFID RF signal might be affected or enhanced by multipath reflection from the side walls, ceiling, and floor. To confirm this suspicion, the test was repeated in

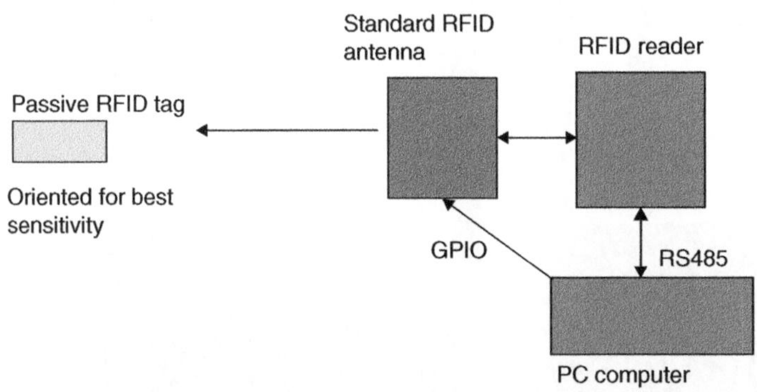

Figure 11.9 Block diagram of distance test setup

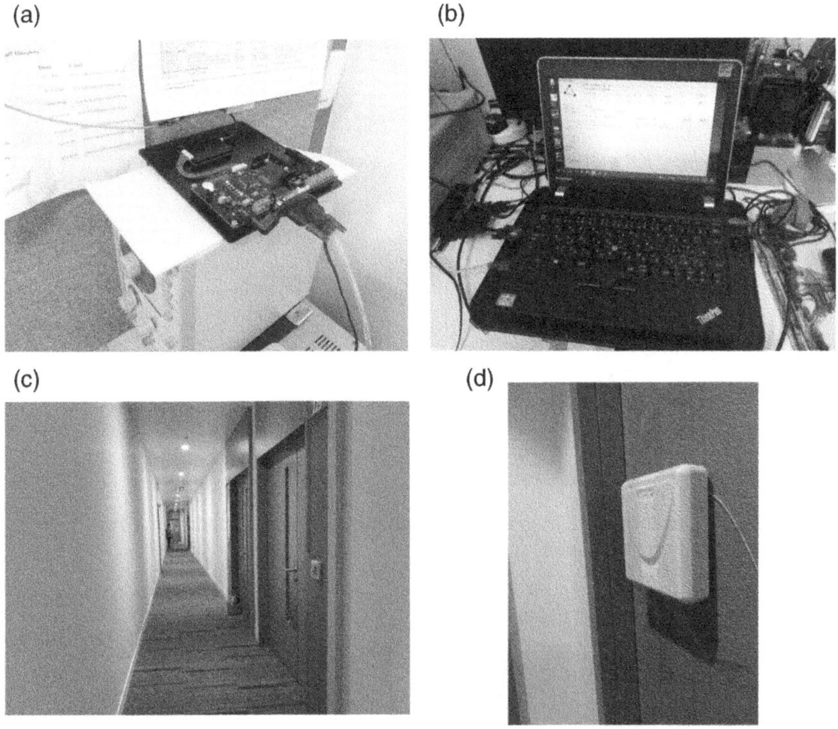

Figure 11.10 Distance test setup: (a) RFID reader, (b) test program, (c) corridor A for testing, and (d) testing patch antenna [Courtesy of Lexiwave Technology (Hong Kong) Limited]

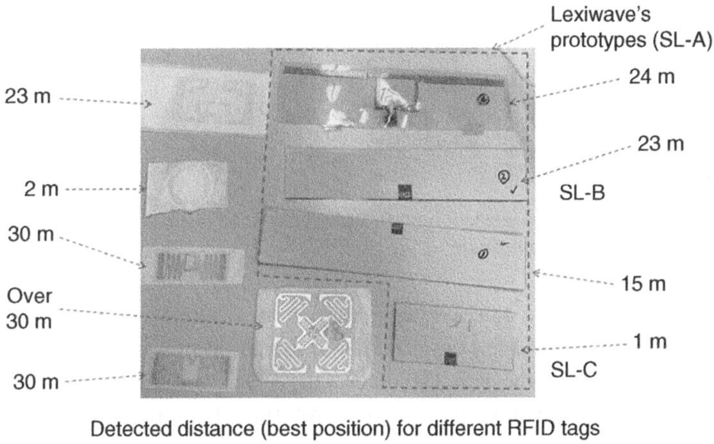

Figure 11.11 Test result: detection distance on passive RFID Tag first prototype

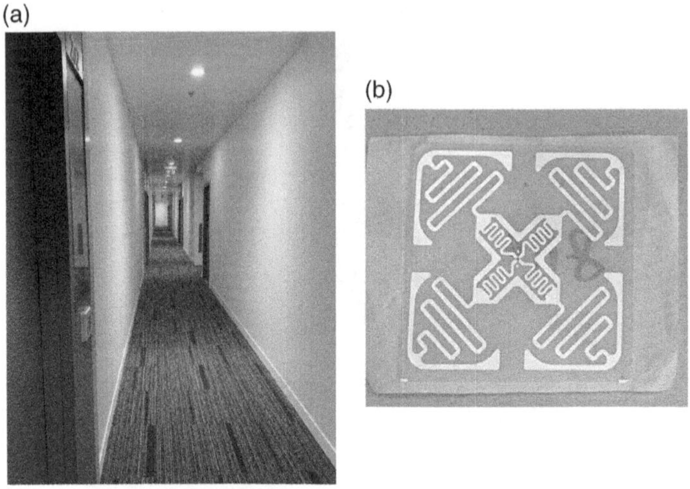

Figure 11.12 Distance test: (a) in Corridor B and (b) detected distance of 40 m for an existing RFID tag on the market

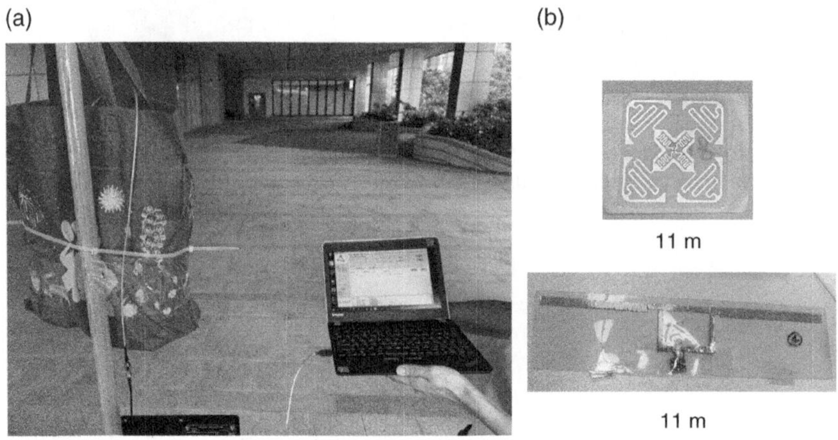

Figure 11.13 Distance test: (a) in outdoor area B, and (b) detected distance drops to 11 m for both standard and designed prototype

the outdoor area at the podium of a building, as shown in Figure 11.13. The measured detection distance dropped to 11 m. In order to further improve the detection range, a newer RFID chip, the NXP SL3S1205 UCODE8 [16], with higher sensitivity was used for designing and making the second prototype. The antenna designed is an inverted-F dipole. The sample was tested with a detected distance of greater than 40 m in Corridor B, and distances of 37 m and greater than 43 m in open areas A and B, respectively (see Figure 11.14).

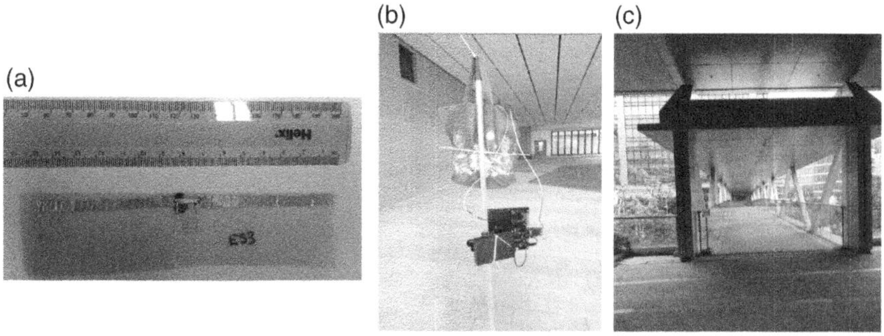

Figure 11.14 (a) Second passive RFID prototype, (b) tested distance >37 m in outdoor area A, and (c) tested distance >40 m in outdoor area B

Despite the long distances measured, open areas A and B might still experience enhanced detection distances due to RF signal reflections from the ceiling and side walls. The detection distance could be shorter in truly open areas without ceilings or side walls. For more accurate measurements, a true open area should be used for the distance test.

Although the detection distance offered by the new prototype was long, its size was quite large (28 mm × 160 mm). It was desirable to make it smaller while maintaining or improving the detection range. Since the chosen RFID chip already had very high detection sensitivity, the only way to further improve the detection range was to enhance the antenna. Typical antenna types include monopole, dipole, and folded dipole, as well as more advanced structures such as fractal and slot antennas [17, 18]. These structures generally offer either good performance or small size, but not both simultaneously. For this application, a more advanced structure, the metamaterial structure [15, 19], was utilized to further enhance the tag's performance. The metamaterial structure is a homogeneous material not found in ordinary materials. It is renowned for its miniature size while offering superior performance under certain conditions.

Two different tags using distinct metamaterial structures [19, 20] were fabricated from regular PCB material (FR-4). The two fabricated tags, Met-A and Met-B, along with other regular structures, are shown in Figure 11.15 for illustration and performance comparison. The same measurement setup as shown in Figure 11.9 was used to measure the receiving sensitivity of the RFID tag at a reference distance of 0.45 m. The numbers shown in Figure 11.15 represent the measured receiving sensitivity of the RFID tag oriented in the optimal position.

The new tag Met-A, featuring a metamaterial antenna structure, offered the best sensitivity (which translates to the longest distance). Although the thickness of the PCB material was only 0.2 mm, it was difficult to bend, limiting its variety of applications, such as luggage tags. A final version using flexi-film material of 0.1 mm was designed and produced, achieving similar performance, as shown in Figure 11.16.

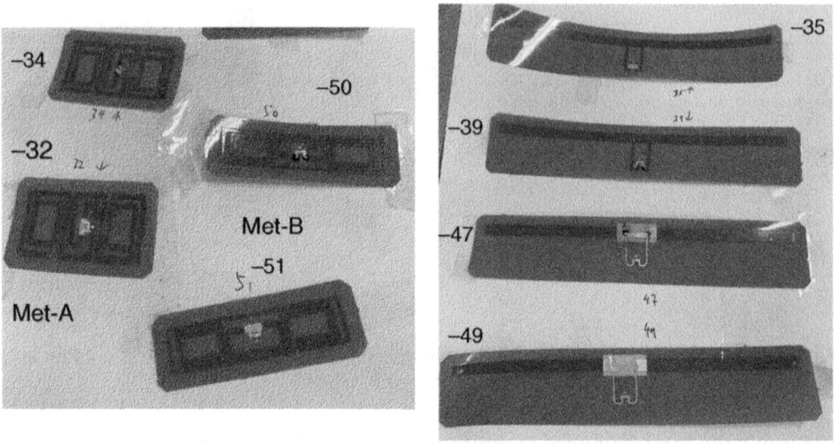

Figure 11.15 Measured signal strength (dBm) at 0.45 m on RFID tags with metamaterial antenna and other regular structures

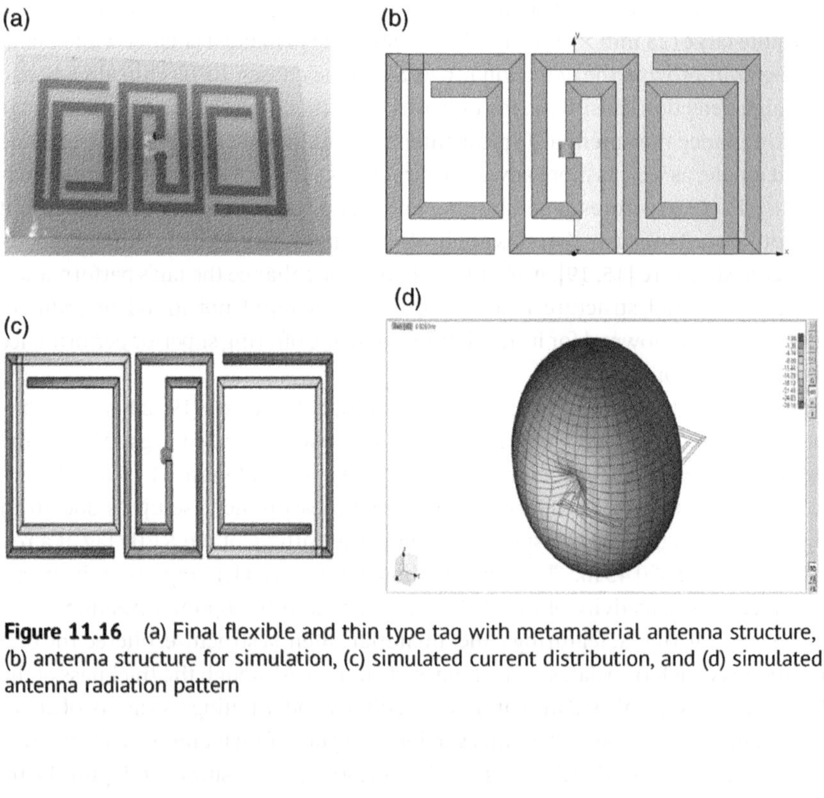

Figure 11.16 (a) Final flexible and thin type tag with metamaterial antenna structure, (b) antenna structure for simulation, (c) simulated current distribution, and (d) simulated antenna radiation pattern

Its read sensitivity was 3 dB better than most passive RFID tags on the market. This improved sensitivity corresponds to about a 50% increase in detection distance. Its flexible and thin structure can be mass-produced for various applications.

To further improve the detection range, a battery-assisted passive (BAP) RFID tag was developed. Two BAP designs were created: the photo-assisted and battery-assisted versions. The BAP RFID chip EM4325 [21] was used for development and evaluation. In the BAP configuration, the EM4325 offers a superior reading range and reliability compared to purely passive RFID solutions. For example, the best read sensitivity of the NXP U8 passive RFID chip mentioned earlier is −23 dBm. With battery assistance, the EM4325 achieves the best read sensitivity of −30 dBm. The 7 dB difference in sensitivity corresponds to more than double the improvement in detection distance if designed properly.

Three different versions of photo-assisted tags with varying antenna structures (metamaterial, meandered dipole, and bent dipole) were created for performance comparison, as shown in Figure 11.17. They were fabricated with FR4 PCB material to provide support for the photodiodes and the RFID chip, as the chip is in package form. The sensitivity of these three versions was suboptimal because the antenna efficiency was relatively low due to an impedance mismatch with the inputs of the RFID chip. The extra connecting traces from the photodiode section to the antenna feed points caused additional impedance mismatch and

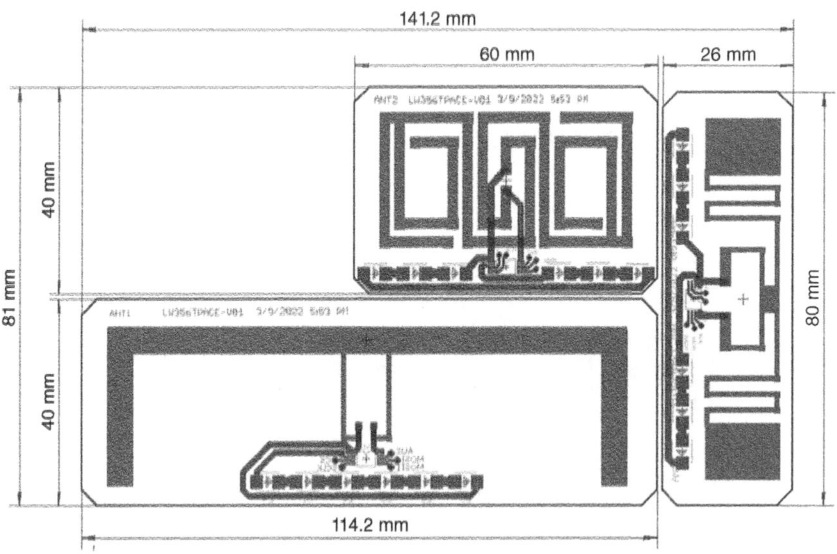

Figure 11.17 Three different versions of photo-assisted tags

further reduced sensitivity. The tag's sensitivity was even lower than that of the developed passive tag with the metamaterial antenna.

It was also found that the output impedance of the photodiode was very high, and the output voltage dropped when the RFID chip tried to transmit its signal to the RFID reader. This voltage drop resulted in poor read sensitivity. To alleviate these problems, a second revision was developed. Although the metamaterial antenna structure offers higher antenna gain, the feeding points are located at the center of the antenna and surrounded by other antenna segments. The feeding points required a long wire in another PCB layer to connect to the inputs of the RFID chip. The extra length of the feeding traces and their overlap with the antenna segments posed significant challenges for impedance matching, as illustrated in Figure 11.18. Therefore, instead of using the metamaterial antenna structure, a regular differential meandered inverted-F antenna was used. This design offered easier matching and required only one layer for connecting the photodiode section to the RFID chip.

The structure of the second revision for simulation using the Antenna Simulator WIPL-D [22] is shown in Figure 11.19, along with the simulation results. The simulated antenna gain was comparable to that of the passive metamaterial antenna tag. The number of photodiodes in this revision has been doubled to lower the output impedance and reduce the voltage drop, ensuring that the reading sensitivity is maintained. The prototype of the second version is shown in Figure 11.20. The read sensitivity in this revision was very similar to that of the passive metamaterial antenna. However, it should be 7 dB better with the power supplied by the

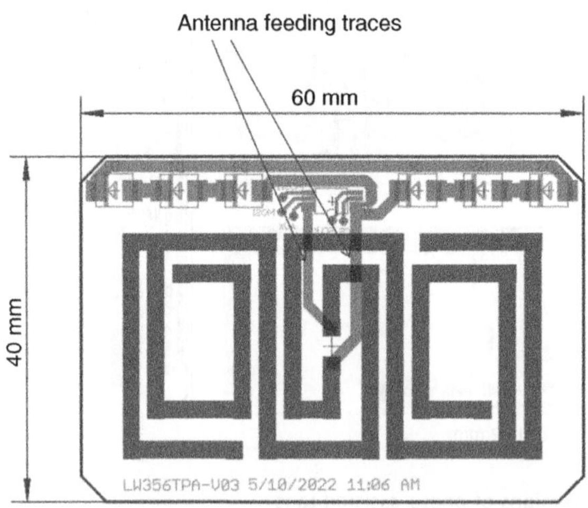

Figure 11.18 Metamaterial antenna structure requiring the second trace layer

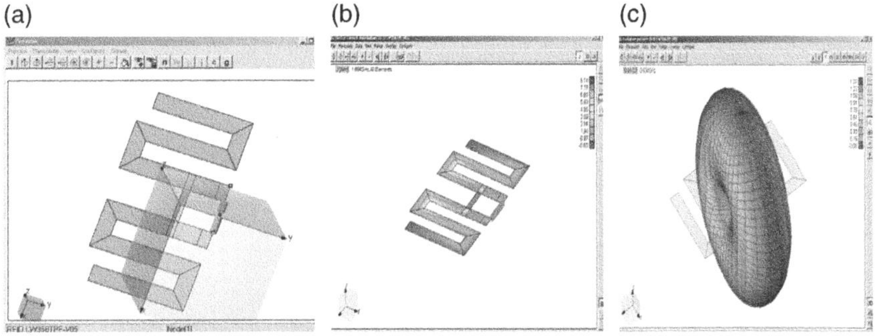

Figure 11.19 (a) Antenna structure for simulation, (b) simulated current distribution, and (c) antenna gain

Figure 11.20 Second revision of the photo-assisted RFID tag

photodiodes. Fine-tuning of the antenna geometry for better impedance matching was performed, and the final prototype is shown in Figure 11.21.

The test results on read sensitivity indicate that this final version had approximately 6 dB higher sensitivity than the passive metamaterial antenna. This improvement was very close to the theoretical difference of 7 dB, as quoted in the datasheet of the EM3245 chip (BAP type) and the NXP SL3S1205 UCODE8 chip (passive type). The prototype was further tested for functionality under different light settings. In a typical office with an LED light tube installed on the

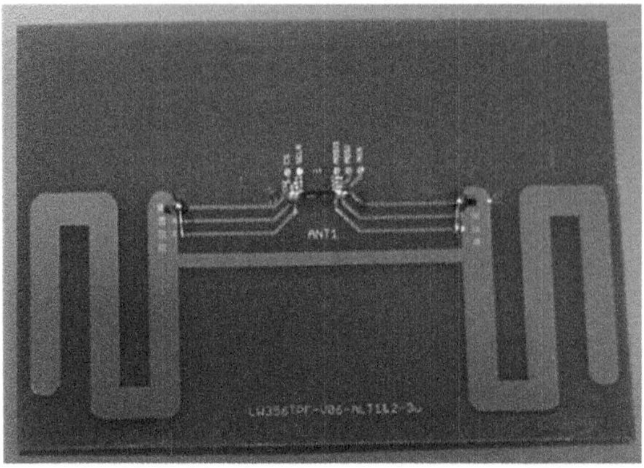

Figure 11.21 Final prototype of the photo-assisted RFID tag

ceiling at around 3.5 m in height, the RFID tag could collect sufficient light energy from the light tube for normal operation, even when the tag was perpendicular to the light source. However, if the ceiling height reached 8 m, as in many industrial buildings or shopping malls, the tag could not function. To alleviate this problem, the tag was modified to include a small rechargeable battery, as shown in Figure 11.22. In this battery-assisted design, the antenna was the same as in the photo-assisted design, but the photodiodes were replaced by a battery. The tested read sensitivity was the same as that of the photo-assisted version.

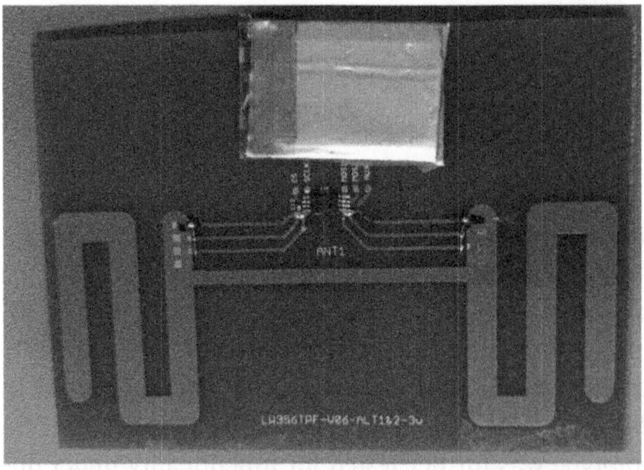

Figure 11.22 Battery-assisted RFID tag

11.4.1.2 Smart Antenna

To extend the RFID tag detection coverage, a smart antenna with beamforming [23] was developed. Its block diagram is shown in Figure 11.23. Three antenna arrays were used for beamforming. The beaming performance was simulated using an antenna array calculator, as shown in Figure 11.24, for different phase shifts in each antenna. With a maximum of 180° provided by the phase shifter, the simulated effective beaming angle is ±50°.

The complete smart antenna comprised a standard RFID reader, three linear phase shifters, three patch antennas, a phase shift controller, and a single-board computer. The linear phase shifter design was chosen over the step phase shifter design because it offered continuous beam sweeping to avoid any blind spots. Each beaming angle provided a unique receiving signal characteristic to the reader from the tag. Therefore, continuous beaming effectively provided an infinite number of virtual readers. When used in localization applications, tracking accuracy could be greatly improved by the smart antenna due to the higher number of datasets collected with continuous beaming. All the constituent parts are shown in the assembly diagram in Figure 11.25.

With the designed RFID tag featuring the metamaterial antenna, the detected distance was around 15 m, as shown in Figure 11.25, which is 3 dB higher than most passive RFID tags on the market. The improved sensitivity corresponds to about a 50% improvement in detection distance. Such an improvement would allow it to be used in indoor location tracking applications, such as inventory

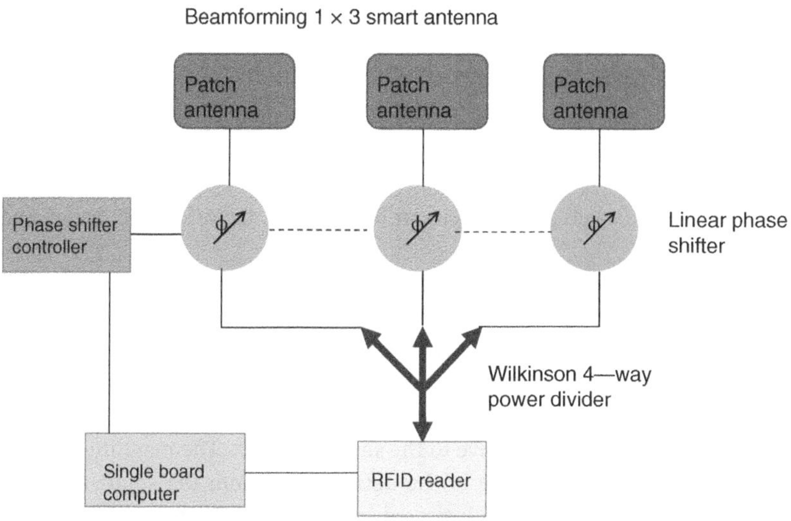

Figure 11.23 Block diagram of the smart antenna with beamforming

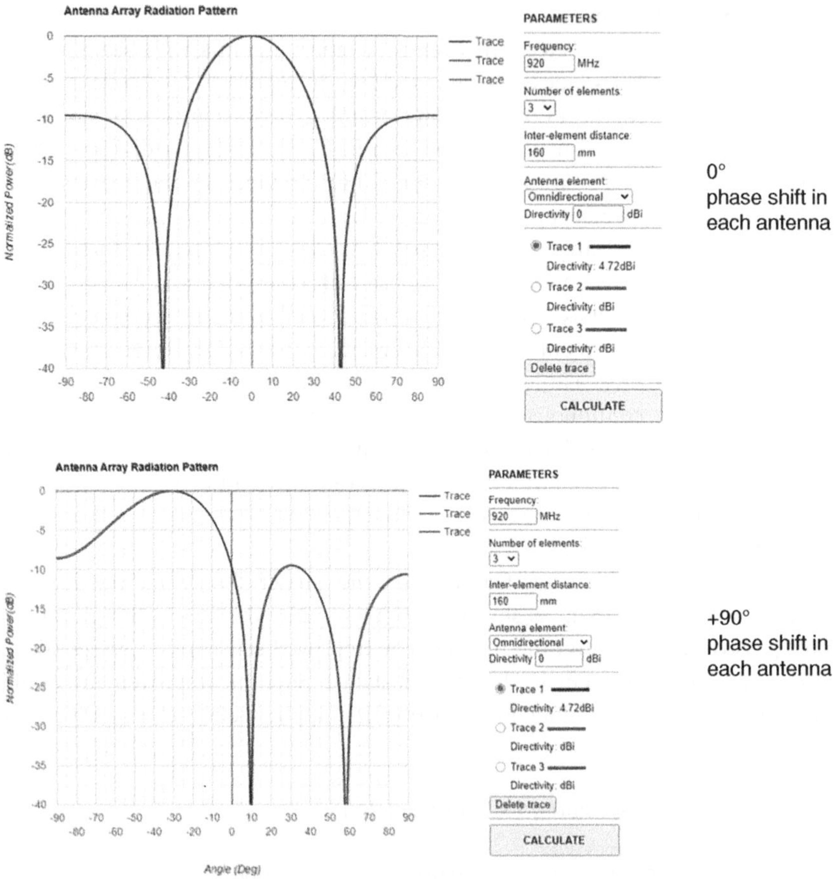

Figure 11.24 Simulated results on the beaming angle with a three-antenna array

and asset tracking, as well as management in hospitals, factories, and designated premises.

The unit was further tested for its beamforming function in location tracking, as shown in Figure 11.26. The detection method used a basic angle detection algorithm. The control program swept the antenna beaming angle from the far left to the far right and recorded the received signal strength Radio Signal Strength Indicator (RSSI) from the tag. The swept angle with the highest RSSI value indicated the direction of the tag relative to the smart antenna. The magnitude of the RSSI value indicated the distance of the detected tag from the smart antenna. The screen on the computer, as illustrated in Figure 11.27, shows the location of the detected tag. The algorithm could successfully determine the angle of the

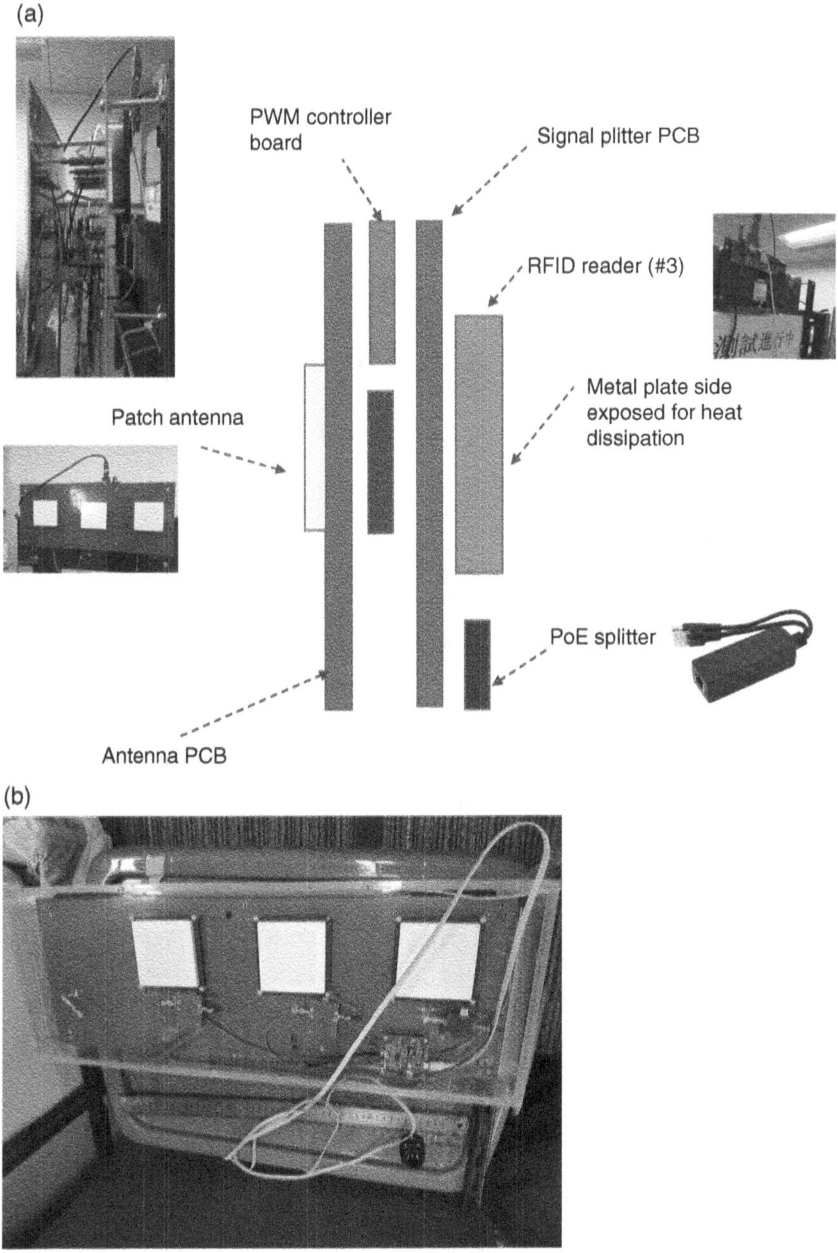

Figure 11.25 (a) Structure of the smart antenna and (b) complete smart antenna

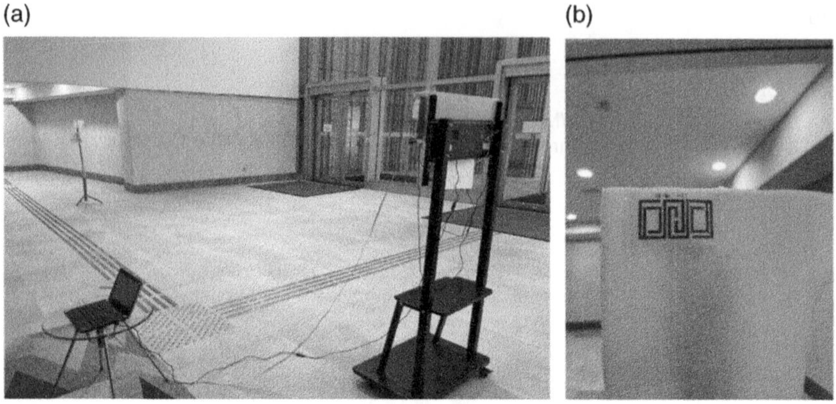

Figure 11.26 (a) Distance detection setup and (b) metamaterial antenna RFID tag for testing

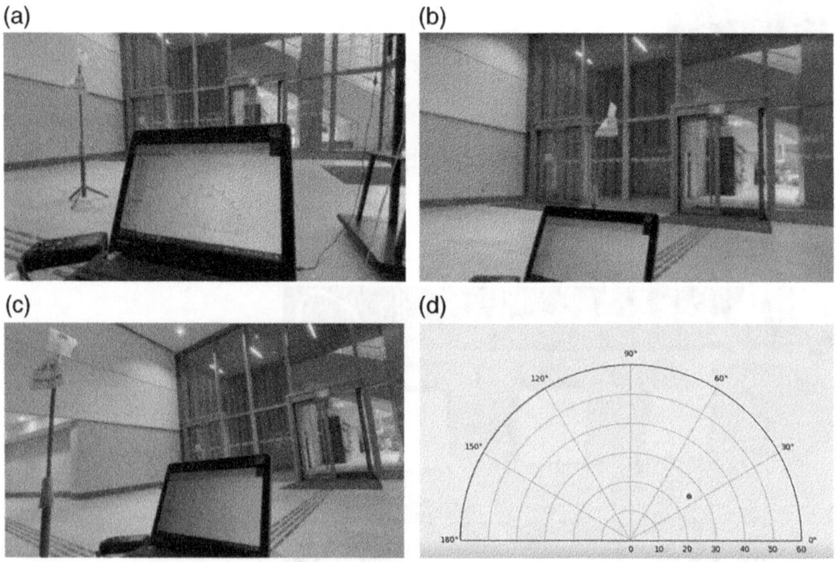

Figure 11.27 Successful location tracking for (a) tag placed in the middle, (b) tag placed to the right, (c) tag placed to the left, and (d) identified location on computer screen

tag with moderate accuracy and could also indicate the relative distance. The absolute distance was affected by several environmental factors, such as signal reflections, the characteristics of the smart antenna, and the tags themselves. Other location methods, such as the Fingerprinting Localization Method, would be needed to achieve higher location tracking accuracy. The beaming angle achieved by the smart antenna is around ±50°.

11.4.2 AI Design

To facilitate the use of AI technology to track RFID tags more precisely, multiple smart antennas are needed to locate and track the tags. The system block diagram of this system is shown in Figure 11.28. The AI computer used was the Nvidia Jetson Xavier NX AI supercomputer for embedded and edge systems. It also served as the web server for data upload and download for the smart antennas, facilitating location tracking.

11.4.3 Indoor Localization

Many studies have been conducted, and their results have been published on indoor localization schemes with satisfactory performance. Among these approaches [24–29], fingerprinting-based (FP) methods are the most widely used due to their effectiveness and simplicity in implementation. FP localization methods are based on the assumption that each location has a unique signal feature. The signal feature is the RSSI value received by each smart antenna from the tag to be tracked. In conventional FP localization systems, multiple RFID readers must be used. However, in the smart antenna developed for this project, the beaming angle is controlled and swept by the controller. The smart antenna can be

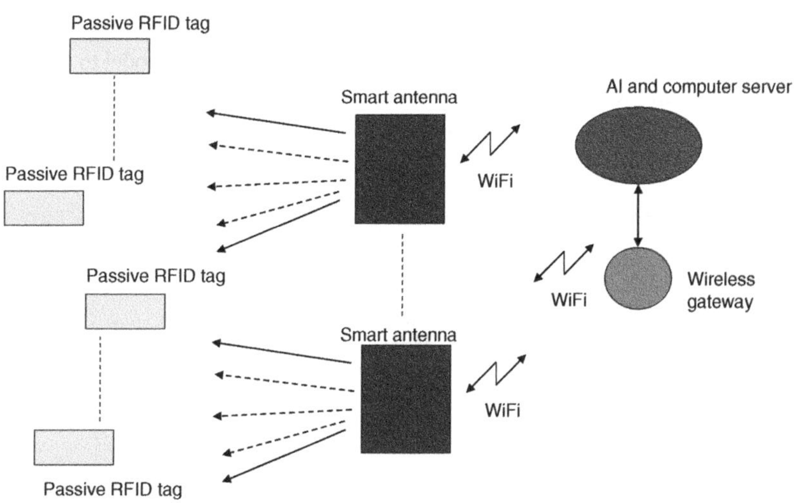

Figure 11.28 System block diagram of indoor tracking with AI technology

considered as multiple virtual readers, as each signal received at each beaming angle is unique and independent.

The FP process involves two steps: (a) offline training and (b) online processing. In the offline step, the RSSI at different reference points (RPs) is collected to construct a radio map. In the online step, the RSSI is sampled at the tag's current position, and the software algorithm finds the most similar signal pattern in the database; the associated RP is then the predicted location.

11.4.4 AI Algorithm

The AI algorithm used is a deep learning (DL) method for FP localization. One of the DL techniques, the CNN, was chosen due to its robustness and popularity in DL architecture. It can automatically detect important features without any human supervision. CNN is typically used for object detection, and in this application, the object is the zone number. The CNN-FP algorithm consists of three parts: (a) data collection, (b) AI model building, and (c) testing and verification (see Figure 11.29).

For data collection, the collected RSSIs from each beaming angle and each RFID reader are unique. This uniqueness forms the basis of the features for the AI algorithm to identify. The AI program converts the collected RSSI values into grayscale values with normalization between 0 and 1 (where darker means weaker RSSI). It then generates a grayscale image matrix based on the grayscale values, the RFID readers, and the beaming angles (see Figure 11.30). The AI algorithm then uses the image classification CNN model to classify the zones. The ResNet-18 [30] model is used to train the model with classes corresponding to the zone numbers and their respective RSSI images. The model generated after optimization will contain the features of the zone to be detected. The AI program then uses the model to predict the zone based on the images it learned from the training. The complete process of AI data collection, training, and verification is illustrated in Figure 11.31.

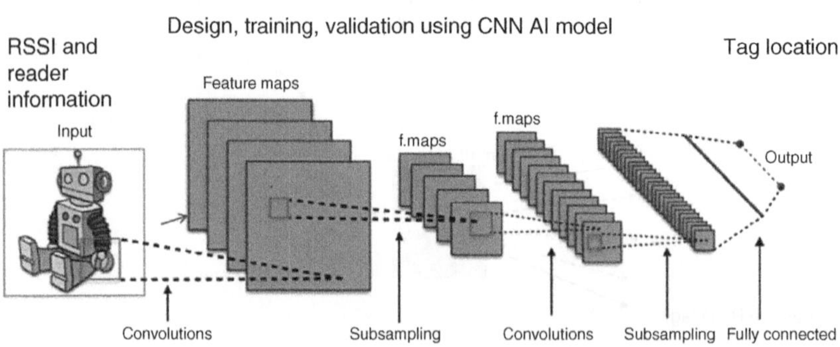

Figure 11.29 Location tracking using CNN AI model for fingerprinting localization [Aphex34/Wikipedia/https://en.wikipedia.org/wiki/File:Typical_cnn.png/CC BY 4.0]

Sweep angle	Reader A gray scale image converting from RSSI value	Reader B gray scale image converting from RSSI value
–60		
–56		
–48		
–40		
–32		
–24		
–16		
–8		
0		
+8		
+16		
+24		
+32		
+40		
+48		
+56		
+60		

Figure 11.30 Image generation for AI classification

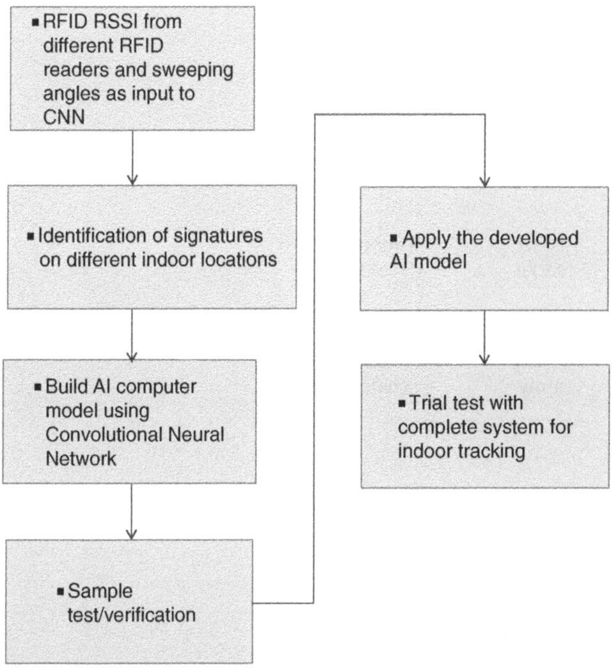

Figure 11.31 Using CNN algorithm to detect the indoor location of the RFID tag by fingerprinting localization technology

11.4.5 Localization Testing and Verification

A preliminary test system with CNN AI was built, as shown in Figure 11.32. The test site was an office space divided into nine zones, each with an area of 1 m × 1 m, as shown in Figure 11.33. Initially, only one smart antenna was used for data collection and AI model training. An off-the-shelf RFID tag was used for the test, as shown in Figure 11.34. The tag was mounted on a wooden stick at a height equal to that of the smart antenna, which was 1.5 m. The tag was placed in each zone, and the RSSI received by each reader was recorded and sent to the AI computer for dataset creation. The tag was then rotated to four angles: 0, +90, +180, and +270°, and the corresponding data were all recorded. The dataset was split 80-15-5 for training, validation, and testing, respectively. Each zone represents one class for the trained model to classify.

After the CNN AI model was built, the predicted tracking accuracy (the probability of predicting the zone number correctly by the CNN AI model) was around 80%. The result was only moderate, as some of the extreme corners may not have been detected by the reader. Therefore, an additional reader was added, and they were placed face-to-face, as shown in Figure 11.33, to cover all the zones. The tracking accuracy increased to 90%. Field verification tests showed that the tag's

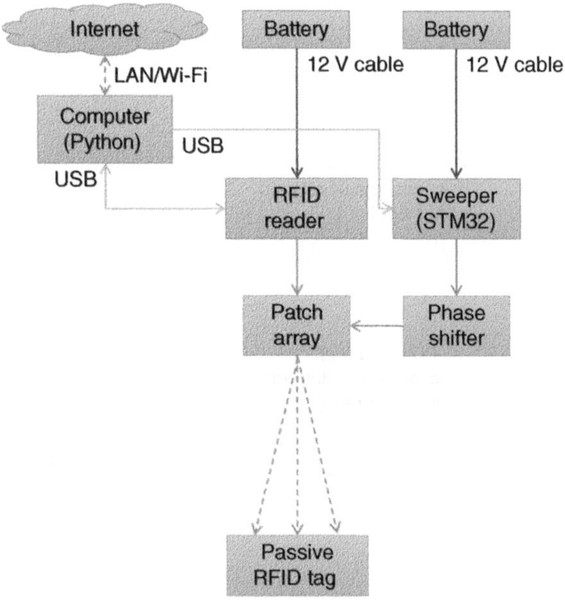

Figure 11.32 System block diagram of indoor tracking with AI technology

(a)

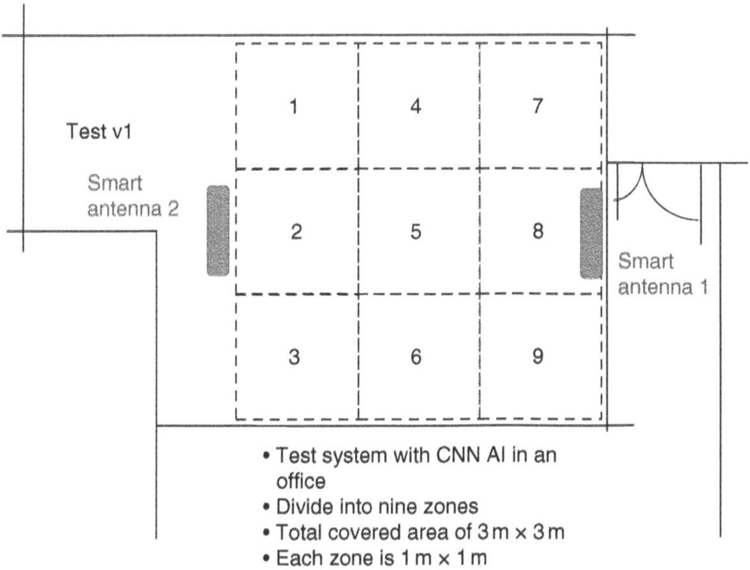

(b)

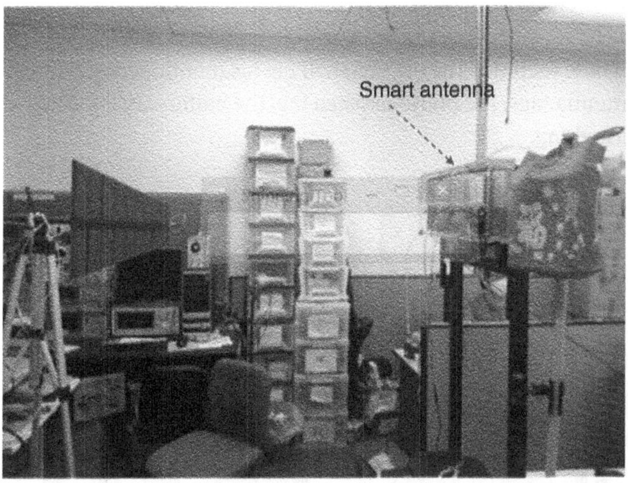

Figure 11.33 Data collection for CNN-FP in an office: (a) dividing the area into zones for tracking and (b) the smart antenna for tracking the RFID tag

location could be identified correctly if the RSSI level at that location was high enough. However, if the tag was located near the edge of the tracking area, the accuracy decreased due to the weak signal level and reflections from the walls and furniture in the office.

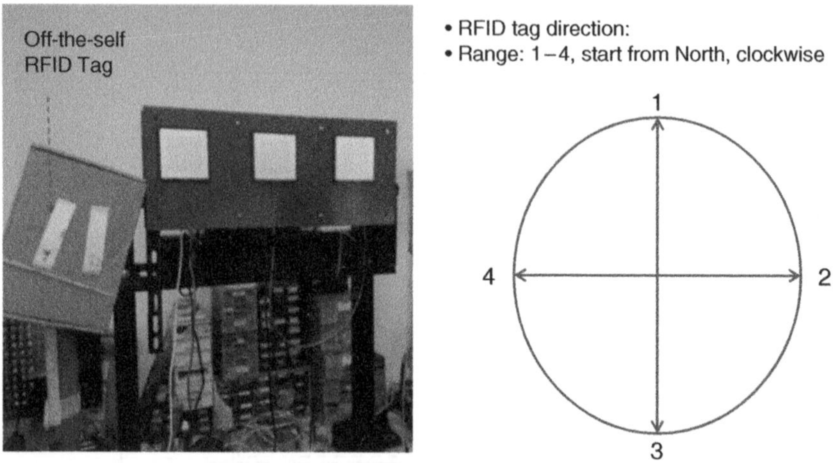

Figure 11.34 Off-the-shelf RFID Tag for testing and rotated for four different angles

After the initial location evaluation, a final test was conducted in a larger indoor test setting with a higher ceiling, as shown in Figure 11.35. For performance comparison, several different RFID tags were placed on a luggage bag for testing, as shown in Figure 11.36, since one of the potential applications is luggage tracking inside airports. These included the off-the-shelf tag (as mentioned previously), two metamaterial antenna tags, the photo-assisted tag, the battery-assisted tag, and the photo sheet-assisted tag.

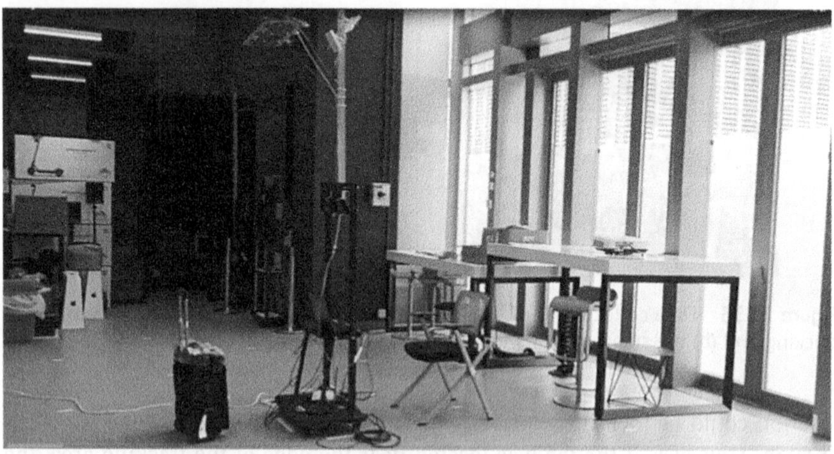

Figure 11.35 A larger test site

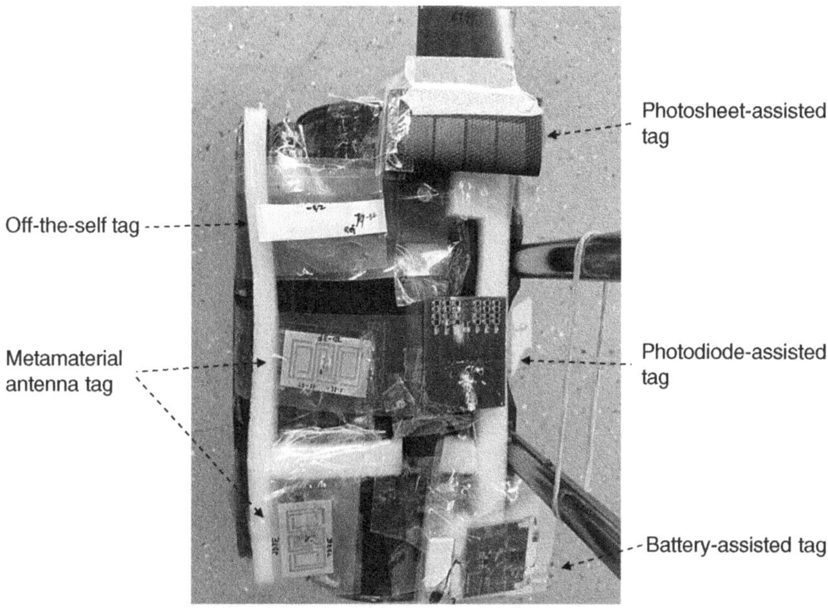

Figure 11.36 Various RFID tags placed on top of a luggage bag

As illustrated in the tracking tests mentioned in previous sections, high tracking accuracy requires high detection signal levels. Another issue that might degrade tracking capability is the radiation pattern of the RFID tag being used. Most planar RFID tags on the market have similar donut-shaped radiation patterns due to their planar structure (see Figure 11.37). The directions shown indicate the blind spots where the lowest transmission and receiving signals are experienced. Since the tag can be placed in any position and orientation with respect to the smart antenna, it might not be detected at certain beaming angles due to these blind spots.

To reduce the effect of blind spots that could potentially degrade tracking accuracy and capability, a new RFID reader from Impinj [31] with higher performance was used. It features four RF channels with a maximum receiving sensitivity of −92 dBm, while the previous model had only −85 dBm. During testing, it was found that the bottleneck for tag detection was the receiving sensitivity of the RFID reader, as the sensitivity of the tags continued to improve. The antenna array was redesigned with larger patch antenna elements, which have a higher overall antenna gain, as shown in Figure 11.38. In addition, two small antennas were added as auxiliary antennas to help track the tag between the left and right positions, as shown in Figure 11.39. These antennas were driven by the second and third RF channels of the new RFID reader.

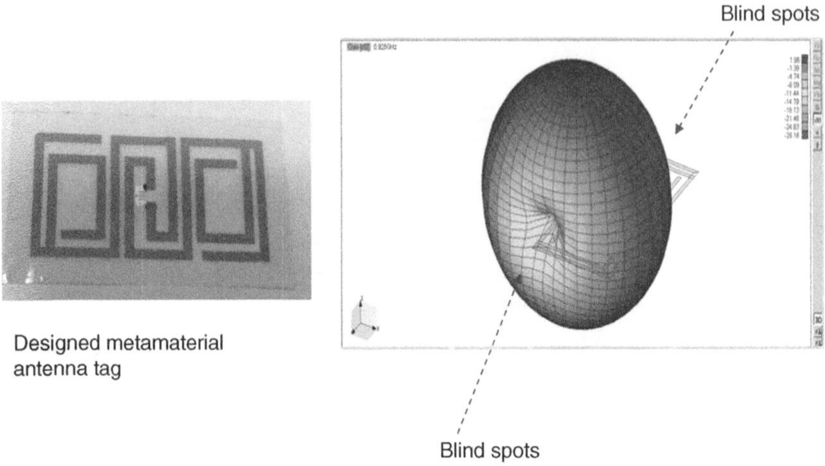

Figure 11.37 Illustration of blind spots in a typical planar antenna

Figure 11.38 Improved antenna array

All the tags were tested to determine the detection distance for the optimal position. The maximum distances were around 25 m for the photo-assisted and battery-assisted tags, while the metamaterial antenna type achieved around 20 m. The off-the-shelf tag had a detection distance of only 15 m.

Since the metamaterial antenna tag did not require any external power source and had a good detection distance, it was used for further tracking tests. The zoning arrangement is shown in Figure 11.40, where the smart antenna was placed at

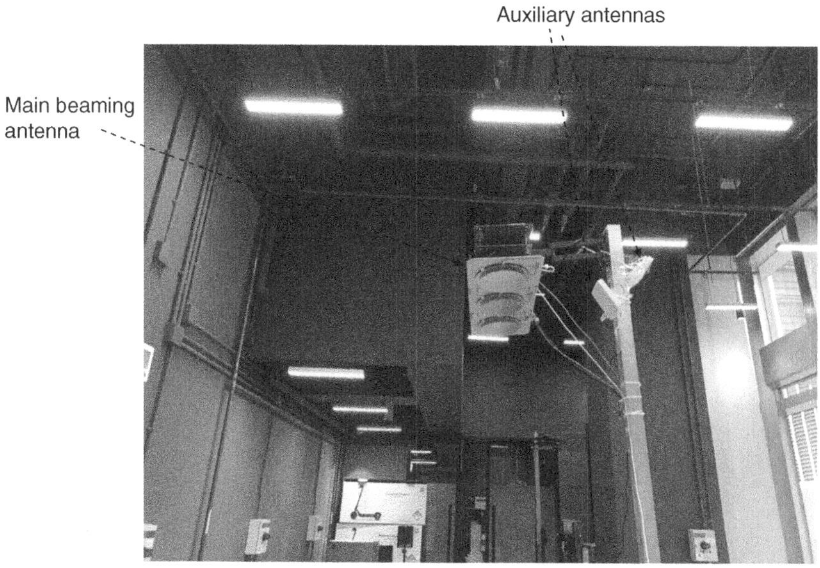

Figure 11.39 High-gain antenna array with two small auxiliary antennas (to differentiate left and right location of the tag)

the center (zone 14). The predicted accuracy of the model was greater than 92%, as the RSSI was strong enough in each zone due to the improved performance provided by the new RFID reader and antenna array. The tag could be detected in all the zones, and the total detected area was larger than 4 m × 9 m.

Figure 11.40 Zoning arrangement for testing with the new reader and improved antenna array

However, for the actual field verification test, the zones near the back side—i.e., zones 1, 7, and 12—could be predicted as the opposite side by the AI algorithm. It was found that there was ambiguity in the beaming response of the new antenna array. Since the antenna array elements were larger, the separation between them had to be greater than 230 mm, which was longer than the optimum distance of 160 mm. As shown in the simulated antenna array radiation pattern (see Figure 11.41), the signal level at the main beaming angle was very close to that at the extreme angles. This ambiguity or symmetry along the beam edges would confuse the AI model, resulting in the predicted zone being on the opposite side. The

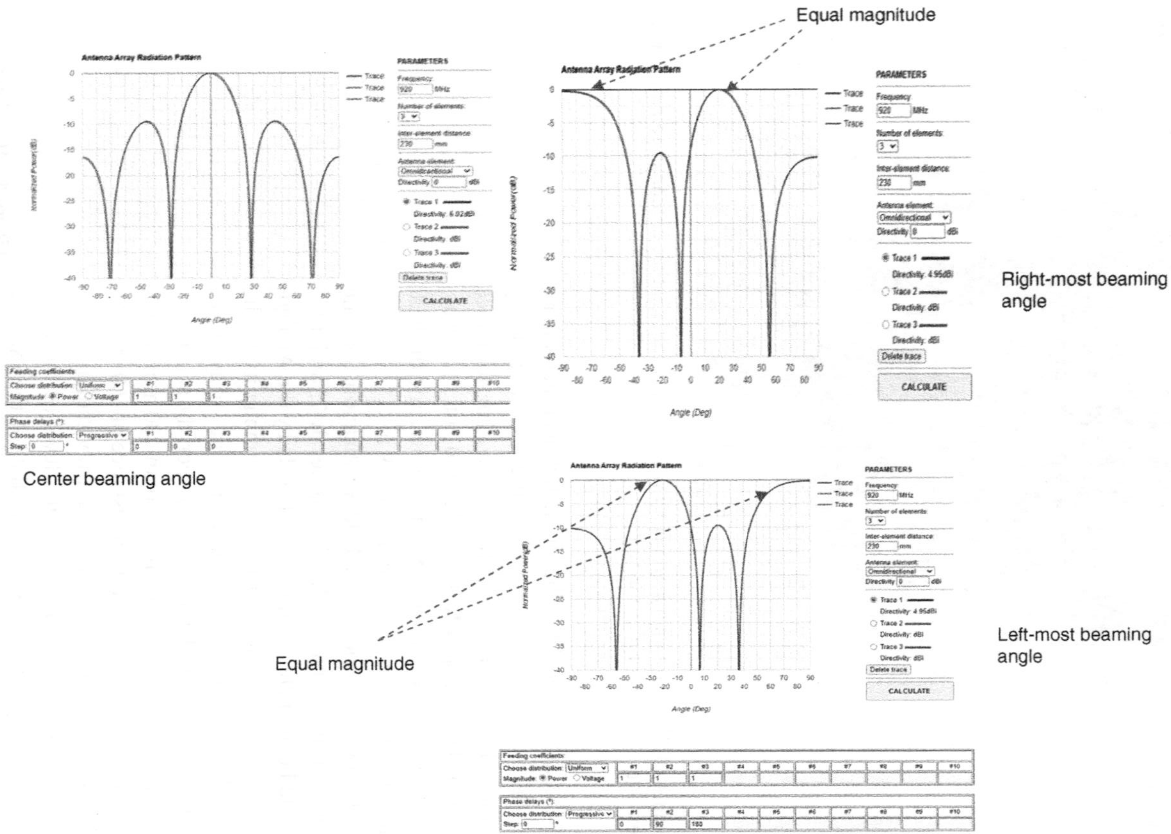

Figure 11.41 Zoning arrangement for testing with a new reader and an improved antenna array

problem could be resolved automatically if an additional smart antenna were used, as the extra RSSI collected from the second antenna would provide asymmetry in the RSSI data pattern. The location could then be identified correctly due to the additional RSSI information. For actual deployment, multiple smart antennas would be used, and the problem would not exist.

Another method is to add a small auxiliary antenna pointing to the back or front side. This creates asymmetry in the collected RSSI pattern, avoiding the ambiguity caused by the beaming response of the large antenna array elements.

References

1 Javed, A.R. et al. (2022). "Future smart cities: requirements, emerging technologies, applications, challenges, and future aspects," *Cities* 129.

2 "What is IoT?," Oracle, last modified February 17, 2023, https://www.oracle.com/hk/internet-of-things/what-is-iot/#iot-applications

3 PlantWeb University, "Wireless Topologies," Wireless 105.

4 Colbach, G. (2018). *RFID Handbook: Technology, Applications, Security and Privacy*. Amazon.com Services LLC.

5 "What Is LoRa®?," Semtech LoRa Technology, last modified February 29, 2024, https://www.semtech.com/lora

6 "Learn About Batteries," Battery University, last modified February 29, 2024, https://batteryuniversity.com/articles

7 Chang, W.-Y. "The State of Charge Estimating Methods for Battery: A Review,"(2013). *Applied Mathematics*. Hindawi Publishing Corporation.

8 "STM32 Mainstream MCUs," STMicroelectronics, last modified February 29, 2024, https://www.st.com/en/microcontrollers-microprocessors/stm32-mainstream-mcus.html

9 "Raspberry Pi Computers," Product Home Page, last modified February 29, 2024, https://www.raspberrypi.com/products/

10 "Beagleboard," Product Home Page, last modified February 29, 2024, https://www.beagleboard.org/

11 AcSip Technology Corp, "S78S LoRa Wireless Communication Module," Datasheet, 2016.

12 "STM32CubeIDE Product Overview," STMicroelectronics, last modified February 29, 2024, https://www.st.com/en/development-tools/stm32cubeide.html

13 Hillar, G.C. (2017). *MQTT Essentials—A Lightweight IoT Protocol: Send and Receive Messages with the MQTT Protocol for Your IoT Solutions*, 1e. Packt Publishing.

14 NXP Semiconductors. (2018). "SL3S4011 UCODE RFID Chip," Product Datasheet.

15 Lau, H. (2019). *Practical Antenna Design for Wireless Products*. Artech House.

16 NXP Semiconductors. (2021). "SL3S1205 UCODE8 RFID Chip," Product Datasheet.

17 Balanis, C.A. (1997). *Antenna Theory - Analysis and Design*. John Wiley & Sons Inc.

18 Gu, C. et al. (2017). "Dual-band electronically beam-switched antenna using slot active frequency selective surface," *IEEE Transactions on Antennas and Propagation* 65 (3): 1393–1398.

19 Ennajih, A. et al. (2020). "A wearable UHF RFID tag antenna-based metamaterial for biomedical applications," *Bulletin of Electrical Engineering and Informatics* 9 (2): 676–684.

20 Ennajih, A. et al. (2019). "A novel design of miniaturized tag antenna loaded metamaterial spiral split ring resonator," *International Journal of Intelligent Engineering and Systems* 12 (2).

21 EM Microelectronic. (2017). "EM4325," Product Datasheet.

22 WIPL-D Electromagnetic Simulation Software, "WIPL-D Pro," Product home page, https://www.wipl-d.com/index.php

23 Mailloux, R.J. (2015). *Phased Array Antenna Handbook*. Artech House.

24 Sinha, R.S. and Hwang, S.-H. (2019). "Comparison of CNN applications for RSSI-based fingerprint indoor localization," *Electronics*.

25 J. Mier, et al. (2019). "At a Glance: Indoor Positioning Systems Technologies and Their Applications Areas," *Proceedings of the International Conference on Information Technology & Systems*.

26 Youssef, M, Agrawala, A. (2005). "The Horus WLAN Location Determination System," *Proceedings of the International Conference on Mobile Systems, Applications, and Services*, pp. 205–218.

27 Perera, C. et al. (2019). "Contextual location in the home using bluetooth beacons," *IEEE Systems Journal* 13 (3).

28 Liu, T., Shuai, Z. (2018). "Self-Positioning system and algorithm in wireless sensor networks," *Proceedings of the 2018 3rd International Workshop on Materials Engineering and Computer Sciences*.

29 Choosaksakunwiboon, S. et al. (2018). "A pre-processing technique for BLE-based indoor localization," *Proceedings of the 12th International Convention on Rehabilitation Engineering and Assistive Technology*.

30 "Transfer learning with PyTorch," last modified February 29, https://github.com/dusty-nv/jetson-inference/blob/master/docs/pytorch-transfer-learning.md

31 Impinj R700 RAIN RFID Reader Datasheet V4_1. Impinj, Inc, 2021.

12

Future Trends

12.1 Introduction

The modern world is increasingly driven by interconnected digital technologies that are transforming how we live, work, and interact. This chapter will explore some of the key advancements that are being made in distributed computing, cloud infrastructure, blockchain, and artificial intelligence (AI)—and how these technologies are enabling new capabilities and use cases, particularly in the burgeoning field of the Internet of Things (IoT) and wireless technologies.

The fundamental concepts of distributed systems, including distributed computing and the various software components that enable distributed systems, will be explored. Examples of how distributed systems are being leveraged in industry for applications like big data processing, the IoT, cloud computing, and AI will be provided. Cloud computing will be discussed as it is crucial for the IoT, providing the infrastructure, tools, and services to manage data, generate insights, and develop intelligent applications. It is seen to enable seamless communication, centralized data management, and simplified IoT solution development, empowering the IoT's full potential.

The rise of cloud computing will then be explored, as it is crucial for the IoT, providing the infrastructure, tools, and services to manage data, generate insights, and develop intelligent applications. It enables seamless communication, centralized data management, and simplified IoT solution development, empowering the IoT's full potential. Blockchain technology, which is enabling secure, transparent, and decentralized data management, will also be spotlighted. Applying blockchain to IoT is seen to offer benefits like improved security, privacy, and autonomous transactions through smart contracts. Additionally, AI, focusing on Machine Learning and Deep Learning, will be introduced. Different types of AI algorithms will also be discussed, including other emerging AI algorithms like Transformers, Explainable AI, Federated Learning, and Hybrid Architectures,

Introduction to Wireless System Design: From Circuits to Web-based Applications, First Edition.
Henry Lau, Ludy Liu, and Keith Chan.

emphasizing the growing importance of explainability, trust, privacy, and adaptability in AI systems.

Finally, the emerging paradigm of edge computing will be examined, which pushes intelligence and processing out to the "edges" of the network, closer to the IoT devices and sensors generating data. This architecture can be seen to improve latency, efficiency, and autonomy for IoT applications. By the end of this chapter, readers will be expected to have a solid understanding of the key technologies shaping the future of distributed, connected, and intelligent systems—and how these innovations are poised to revolutionize the IoT.

12.2 Distributed System

A distributed system [1] is a collection of autonomous computers or nodes that work together as a unified system as shown in Figure 12.1. These nodes communicate and coordinate with each other to achieve a common goal. In a distributed system, the nodes are physically independent and can be located in different geographic locations and connected through a network. The main feature of a distributed system is that it allows parallel execution of tasks across multiple nodes. The focus is on the overall system architecture and design, including not only computing, but also communication, coordination, fault tolerance, and resource sharing. Distributed systems can include a wide range of applications and services,

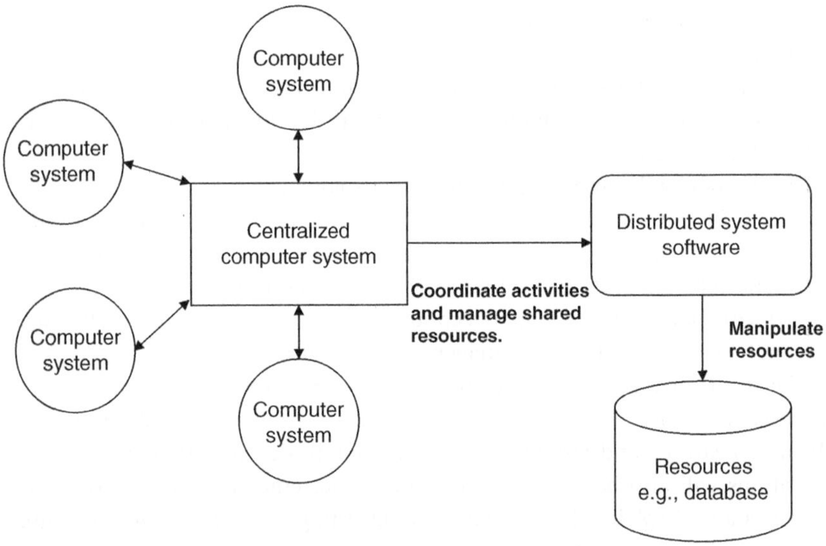

Figure 12.1 Example of a distributed system

including distributed databases, content delivery networks, peer-to-peer networks, and so on. Distributed System Software enables computers to coordinate their activities and share hardware, software, data, and other resources.

12.2.1 Distributed Computing

Distributed computing technology is a branch of computer science that studies how to divide a problem that requires huge computing power into many small parts, then allocate these parts to many computers for processing, and finally combine these calculation results to obtain the final result. It focuses on how the computational workload is divided and assigned to different nodes, and how the nodes communicate and coordinate with each other to accomplish the task. Distributed computing has several advantages over other algorithms. One advantage is the ability to share rare resources. Another advantage is the ability to balance the computing load across multiple computers. Additionally, distributed computing allows for program placement on the most suitable computer. Sharing rare resources and load balancing are core ideas in distributed computing.

12.2.2 Software for Distributed Systems

In a distributed system, various types of software are used to enable communication, coordination, and efficient management of resources across multiple nodes. The following are some of the key software components commonly used in distributed systems:

12.2.2.1 Middleware

Middleware enables certain services that do not exist in the local or centralized system by acting as an interface between the centralized system and the local system. Middleware service can facilitate communication and data management between distributed components. Examples of middleware include Apache Kafka [2].

12.2.2.2 Distributed File Systems

A distributed file system is designed to store and manage files across multiple nodes in a distributed system. Examples include Hadoop Distributed File System [3] and Google's Colossus File System—which is used as the primary storage system for Google services like Google Drive, Gmail, and Google Photos.

12.2.2.3 Distributed Databases

The role of a distributed database in a distributed system is to store and manage data across multiple nodes. It provides fault tolerance, scalability, and efficient data access. Examples include MongoDB [4] and Amazon DynamoDB.

12.2.2.4 Distributed Computing Frameworks

The distributed computing framework supports the execution of computational tasks across multiple nodes in a distributed system. It provides abstractions and tools for task allocation, fault tolerance, and data parallelism. Examples include Apache Spark, Apache Hadoop, and Apache Flink [5].

12.2.2.5 Orchestration Tools

The Orchestration tools support the management and deployment of distributed systems by automatically provisioning, configuring, and monitoring nodes. Examples include Kubernetes [6] and Apache Mesos [7].

12.2.3 Examples of Distributed Applications in Industry

Distributed systems are essential components in a wide range of industrial applications. One prominent example is in the realm of big data processing, where systems like Apache Spark and Apache Hadoop are employed to handle massive datasets. These systems distribute computation and data across clusters of machines, facilitating parallel processing and efficient data storage [8]. Another area where distributed systems are crucial is in the IoT. IoT applications typically involve numerous connected devices generating and processing data. Distributed systems are instrumental in collecting, analyzing, and storing data from these devices, enabling real-time monitoring and decision-making. Cloud computing platforms, such as Microsoft Azure [9], and Google Cloud Platform (GCP), also rely heavily on distributed systems to provide users with virtualized resources, scalable storage, and distributed computing services. Furthermore, distributed systems play a vital role in AI applications. TensorFlow, an open-source machine learning framework, supports distributed computing, allowing for the distribution of training and inference tasks across multiple devices or machines. This capability enables the training of large-scale machine learning models, pushing the boundaries of AI development.

12.3 Cloud Computing

Cloud computing is to provide information, applications, and computing resources to users or organizations (servers, storage, databases, networks, software, analysis, AI, etc.) via the Internet (cloud).

Cloud computing is usually based on virtualization technology [10] to quickly deploy resources or obtain services. Virtualization software simulates physical hardware and runs multiple virtual machines on a single machine. This provides flexibility in interacting with hardware resources. Physical servers consume power, take up storage space, and require maintenance. Virtualization removes these limitations by abstracting physical hardware functions into software. This enables the management, maintenance, and utilization of the hardware infrastructure to be conducted as if it were a web application.

Cloud computing essentially represents the concept of distributed computing as shown in Figure 12.2. It simply means letting a number of different computers do jobs for the user at the same time. No matter how many computers the user has, they can let the computers communicate with each other and do jobs for them at the same time.

12.3.1 Typical Cloud Server

Four distinct cloud storage solutions are available, each with unique characteristics: public, private, hybrid, and multicloud. Public cloud computing architecture is established by vendors selling cloud services to the general public or large industrial groups. Private cloud computing architecture, on the other hand, serves only a single organization and can be managed by the organization itself or by a third-party vendor. It can be deployed both inside and outside the enterprise. Multicloud computing architecture is jointly established by multiple organizations to serve groups with common demands and needs, such as shared missions, security requirements, policy, and regulatory considerations. This type of architecture can be managed by the participating organizations or third-party vendors and can also be deployed inside or outside the organization. Finally, hybrid cloud computing architecture combines a private cloud with a public cloud, integrating them

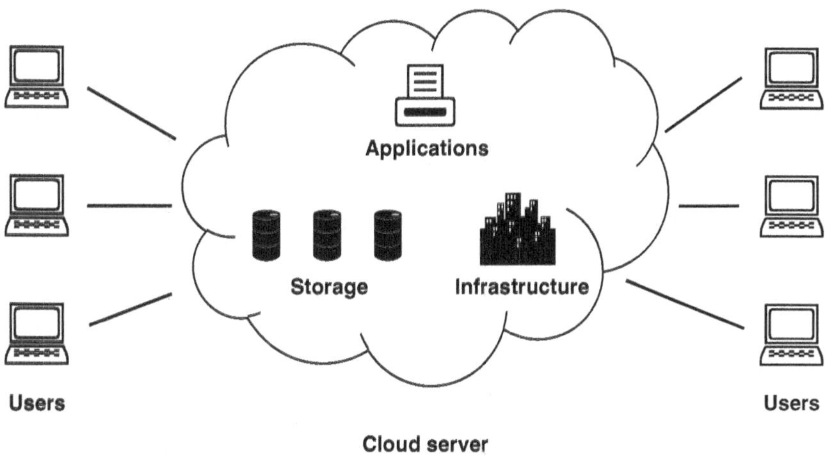

Figure 12.2 Cloud server

through standard or proprietary technologies to allow data portability with applications.

Cloud computing services are also divided into three main types: Infrastructure-as-a-Service (IaaS), Platform-as-a-Service (PaaS), and Software-as-a-Service (SaaS) [11].

12.3.1.1 Infrastructure-as-a-Service

IaaS delivers virtualized computing resources over the Internet. It provides virtual machines, storage, and networking capabilities that allow users to deploy and manage their own operating systems and applications. Customers control the infrastructure layer, while the cloud provider manages the underlying physical infrastructure. Examples of IaaS providers include Amazon Web Services (AWS) EC2, Microsoft Azure, and Google Compute Engine.

12.3.1.2 Platform-as-a-Service

PaaS provides a platform for developing, testing, and deploying applications. It provides a complete development environment that includes infrastructure, runtime, and development tools. With PaaS, developers can focus on writing code without worrying about infrastructure management. The cloud provider manages the underlying infrastructure, including servers, storage, and networking. Examples of PaaS providers include Heroku, Google App Engine, and Microsoft Azure App Service.

12.3.1.3 Software-as-a-Service

SaaS delivers software applications over the Internet on a subscription basis. Users access these applications through a Web browser or thin client, eliminating the need for local installation and maintenance. The cloud provider hosts and manages the entire software application stack, including infrastructure, platform, and software. Examples of SaaS providers include Salesforce, Microsoft Office 365, and Dropbox.

12.4 Role of Cloud Computing in IoT

As discussed in previous Chapters, IoT is a world of interconnected devices, each generating a constant stream of data. The data is recognized as having immense potential for improving efficiency, creating new services, and enhancing lives. However, the managing and harnessing of this data deluge is required to be done through a powerful infrastructure, and cloud computing is seen as a crucial enabler. Cloud platforms, with their vast storage capacity and scalability, are considered the perfect solution for handling the massive amounts of data generated by IoT devices.

As illustrated in the previous design examples on smart city applications, many other applications such as smart traffic lights, sensors monitoring environmental conditions, and countless smart homes, are all sending data to a central location. Cloud storage is used to provide the space to store this data securely and reliably, ensuring that valuable information is not lost or compromised.

Beyond just storing data, the data is empowered to be analyzed by cloud computing. Cloud-based analytics tools are used to sift through the data, identify patterns, and extract meaningful insights. This allows traffic flow to be optimized, maintenance needs for infrastructure to be predicted, or even personalized experiences in smart homes to be created. For example, by analyzing data from smart appliances, a cloud-based platform can learn preferences and automatically adjust settings to optimize energy consumption. Moreover, cloud computing acts as the connective tissue for the IoT, linking devices together and facilitating communication. Imagine a network of sensors in a factory, each monitoring different aspects of production. Cloud platforms are used as a central hub, allowing these sensors to share data, enabling real-time monitoring, and control of the entire production process. This seamless communication is essential for creating truly interconnected and intelligent systems.

The benefits of cloud computing extend to the development and deployment of IoT applications. Cloud platforms offer a range of tools and services that simplify the process of building and deploying these applications. Developers can leverage cloud-based development environments, utilize prebuilt components, and easily scale their applications to accommodate a growing number of devices. This streamlined approach accelerates innovation and allows for the rapid development of new and exciting IoT solutions.

In essence, cloud computing is seen as the bedrock upon which the IoT thrives. It provides the infrastructure, tools, and services necessary to manage the vast amounts of data generated by connected devices, analyze it for valuable insights, and develop intelligent applications that transform our world. From smart homes to industrial automation and smart cities, cloud computing empowers the realization of the full potential of the IoT, creating a future where technology seamlessly integrates with our lives, enhancing experiences and shaping a more connected and intelligent world.

12.5 Blockchain

Blockchain [12] is a decentralized data computing and storage platform achieved by using the Internet. Operating in a decentralized manner, the blockchain stores databases in different computers and monitors each other's data. Data changes require the consent of other computers.

A blockchain is a chain of blocks. Each block contains some information and is linked to the next one in the chain in the order they were created. The chain is saved on all servers, as long as one can work. Nodes in the blockchain system provide storage and computing power for the entire system. The number of nodes needed to reach a consensus varies depending on the algorithm. To change information on the blockchain, one needs to get the consent of at least half of the nodes. These nodes are usually in the hands of different people or organizations, so it is difficult to tamper with the information in the blockchain. The blockchain has two main features: data security and decentralization. The blockchain is more reliable because it is difficult to tamper with the information recorded in it. This can help solve the problem of people mistrusting each other.

12.5.1 Technology Highlights

At its core, blockchain utilizes a decentralized ledger, a distributed database that records transactions across a network of computers. This decentralized approach offers several advantages over traditional centralized systems, making it a powerful tool for secure and transparent data management.

A decentralized ledger [13] means that transaction bookkeeping is done by multiple nodes distributed in different places, and each node records a complete account, so they can all participate in monitoring the legitimacy of the transaction as well as testifying to it together. Compared to traditional decentralized storage, blockchain's decentralized storage has two unique aspects. Firstly, each node in the blockchain stores complete data based on the blockchain structure, while traditional decentralized storage divides data into multiple copies stored according to specific rules. Secondly, in most blockchains, each node in the blockchain storage is independent and has the same state, ensuring consistency through consensus mechanisms. In contrast, traditional decentralized storage relies on a central node to synchronize data with backup nodes. In a blockchain system, no single node can record bookkeeping data alone, preventing the risk of a single bookkeeper being controlled or bribed to manipulate accounts. Additionally, with sufficient nodes recording accounts, the data remains secure unless all nodes are destroyed, ensuring the safety of account data.

Blockchain utilizes advanced, asymmetric cryptographic techniques to ensure the security and privacy of transactions. Each user is assigned a unique pair of keys: a public key and a private key. The public key is used to receive transactions, while the private key is used to authorize transactions. This system ensures that only the rightful owner can access and control their funds. While transaction information is publicly viewable on the blockchain, the identity of the users remains encrypted, protecting their privacy. A consensus mechanism is how a consensus is reached among all accounting nodes to determine the validity of a record, both as a means of authentication and as a means of preventing tampering. Blockchain proposes different consensus mechanisms [14], which apply to different application scenarios and strike a balance between efficiency and security.

12.5.2 Structure of a Block

A blockchain is made up of a series of blocks as shown in Figure 12.3. Each block contains a set of transactions or data and has a fixed size. The block consists of a header and a body. The header contains crucial information, such as a unique identifier called a hash, a timestamp, a reference to the previous block's hash, and additional metadata. The hash is a cryptographic hash function that maintains the integrity and immutability of the block. The body of a block contains the data or transactions being recorded on the blockchain. The content may vary depending on the specific use case of the blockchain. In a cryptocurrency blockchain, the body contains transaction details, such as the sender, recipient, and amount.

Each block in the blockchain links to the previous block through its hash, creating a chain of blocks where the hash of each block depends on the data and hash of the previous block as shown in Figure 12.4. This linking ensures the integrity of the blockchain and makes it difficult to tamper with past blocks.

12.5.3 Leverage Blockchain Technology for IoT Devices

Blockchain technology applied to the IoT has the potential to revolutionize the way devices and systems interact and transact with each other. By combining IoT with blockchain, several challenges associated with security, privacy, data integrity, and trusted transactions can be addressed.

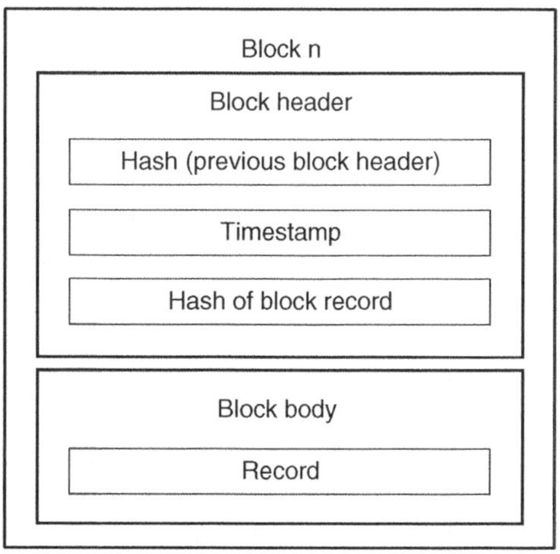

Figure 12.3 Structure of a block

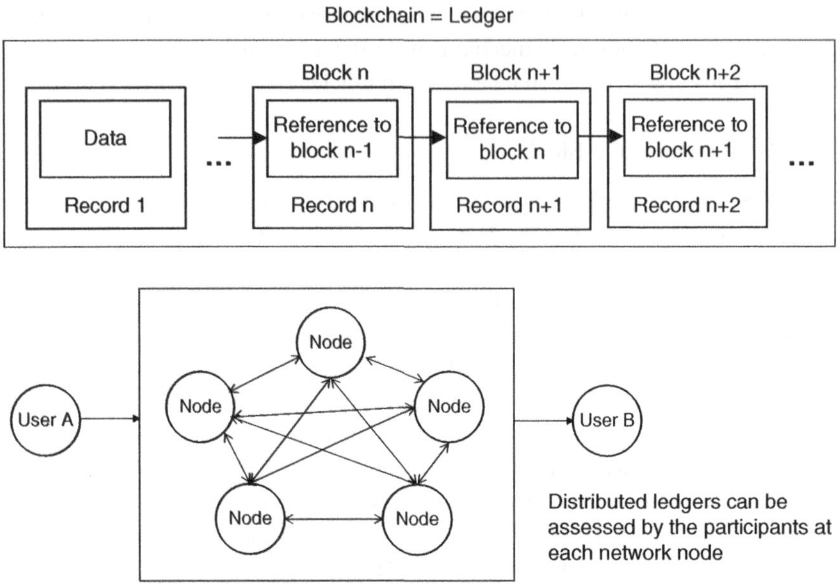

Figure 12.4 Operation principle of a blockchain

The IoT refers to the network of interconnected physical devices, vehicles, appliances, and other objects that are embedded with sensors, software, and connectivity, allowing them to collect and exchange data. Blockchain technology offers several benefits in this context.

Firstly, blockchain provides a decentralized and tamper-resistant ledger where transactions and data exchanges can be securely recorded. The decentralized nature of blockchain makes it difficult for malicious actors to tamper with the data, ensuring the integrity and authenticity of the information being exchanged between IoT devices. Additionally, blockchain technology enables trust and transparency between IoT devices that may not have any prior relationship or knowledge of each other. By recording transactions on a shared and immutable ledger, all participants can verify and trust the information exchanged between devices without relying on a central authority. This can be particularly useful in scenarios where devices need to interact and exchange information autonomously, such as in smart contracts or supply chain management.

Smart contracts, which are self-executing contracts with predefined rules and conditions encoded within the blockchain, can automate payments between IoT devices based on predefined conditions. For example, a smart contract can automatically trigger a payment from one IoT device to another when specific conditions are met, eliminating the need for manual intervention.

Furthermore, blockchain applied to the IoT allows individuals to gain more control over their personal data. They can selectively grant access to their data, specify how it can be used, and even monetize it directly by establishing transparent and auditable transactions. This empowers individuals to retain ownership and benefit from the data generated by their IoT devices. Scalability is a challenge for blockchain technology when it comes to handling a large number of transactions. However, advancements such as off-chain transactions and sharding are being explored to improve the scalability and efficiency of blockchain networks, making them suitable for handling the massive data streams generated by IoT devices.

Overall, the application of blockchain technology to the IoT holds great promise for enhancing security, privacy, trust, and transparency in the increasingly interconnected world of devices. While there are still technical challenges to overcome and practical implementations to be developed, the combination of blockchain and IoT has the potential to unlock new possibilities and enable innovative use cases across various industries. Additionally, the use of blockchain in IoT systems enhances security by ensuring that the attribute information of each IoT device remains hidden while providing a means for administrators to update and verify device configurations.

12.6 Artificial Intelligence

AI is a technology that simulates human intelligence, enabling computer systems to process, analyze, and make decisions on data autonomously through machine learning, deep learning, natural language processing (NLP), and other algorithms. Since the concept of AI was first introduced in the 1950s, AI technology has undergone several waves of development and achieved remarkable progress and results, which have had a profound impact on the global economy, society, and culture.

12.6.1 Machine Learning, Deep Learning, and Artificial Intelligence

Machine learning studies and builds a collection of algorithms that enable a computer to learn from data itself and make predictions, rather than relying on a specific pre-programmed algorithm. Machine learning encompasses many different techniques [15], and deep learning is one of them. Other methods include decision trees [16], clustering [17], Bayesian [18], etc.

Deep Learning is a subset of Machine Learning and refers to artificial neural networks (ANNs) modeled by algorithms that can work like a human brain. Deep learning is inspired by the structure and function of the brain, which is the interconnection of many neurons. ANNs [19] are algorithms that mimic the biological structure of the brain.

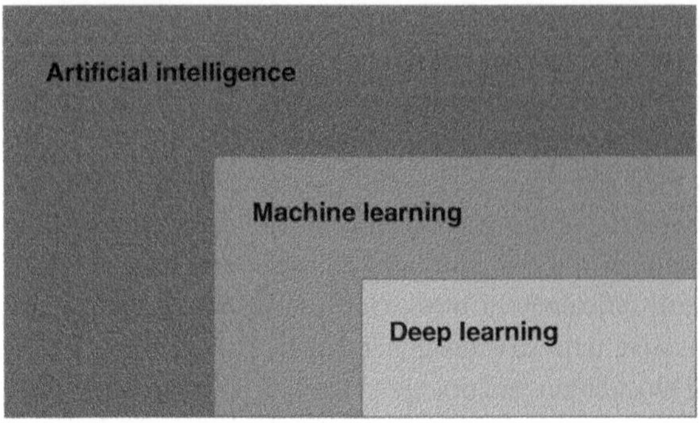

Figure 12.5 AI category

Both machine learning and deep learning belong to the category of AI. AI, machine learning, and deep learning can be represented by the diagram in Figure 12.5.

12.6.2 Machine Learning

Machine learning is a form of AI that lets computers learn from data and improve their performance over time. It involves training an algorithm on a subset of data and using it to predict and make judgments about other data. At its core, machine learning is about algorithms that parse data, learn from it, and make decisions or predictions about new data (see Figure 12.6). A computer uses data to create a model and then makes predictions. This is similar to how humans learn. For example, people can predict new problems after gaining experience.

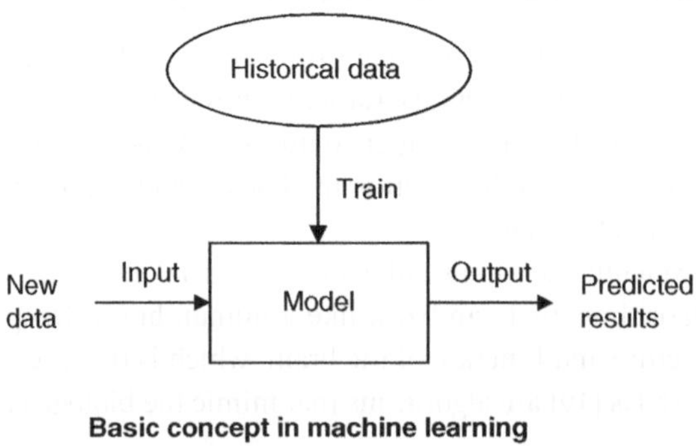

Figure 12.6 Basic concept in machine learning

12.6.3 Machine Learning Algorithms

Machine learning Algorithms are used in a wide range of applications, from image recognition, NLP, and risk detection to self-driving cars.

There are many different approaches including supervised learning, unsupervised learning, semisupervised learning, and reinforcement learning.

Supervised learning algorithms learn from labeled data where the required output for each input is known. The algorithm uses this labeled data to learn how to predict the outputs of new, unseen inputs. Examples of supervised learning algorithms include decision trees, random forests, and neural networks.

Unsupervised learning algorithms learn from unlabeled data, where the desired output is unknown. These algorithms are used to discover patterns and structures within the data, such as grouping similar data points together through techniques like clustering. Examples of unsupervised learning algorithms include k-means clustering [20] and principal component analysis (PCA).

Semisupervised learning algorithms are a combination of supervised and unsupervised learning, where the algorithm learns from both labeled and unlabeled data. This method is often used when labeled data is scarce or expensive to obtain.

Reinforcement learning algorithms learn through repeated trials by receiving feedback in the form of rewards or penalties for each action taken. These algorithms are used in applications such as games, robots, and self-driving cars.

Overall, machine learning algorithms have the potential to revolutionize many industries, making our lives easier and more efficient. However, they also raise important ethical and social questions, such as how to ensure fairness, transparency, and accountability in their use.

12.6.4 Deep Learning

In the IOT field, many sensors send data such as video, sound, temperature, etc. to the cloud server, and it is quite suitable to use Deep Learning for analysis.

Deep learning involves the automatic generation of multiple layers of abstraction during the training of a function. Each layer in a deep neural network has neurons that implement relatively simple mathematical operations, but the combination of hundreds or thousands of these functions forms a powerful model capable of learning complex patterns in data. This branch of machine learning utilizes multilayered ANNs, representative of such as Convolutional Neural Networks (CNNs) and Recurrent Neural Networks (RNNs).

CNNs are more suitable for spatial data types such as pictures and videos and can recognize images through different levels of features, such as a nose feature, an eye feature, a mouth feature, and the relationship between the three, and finally a human face. RNNs are more suitable for sequential data applications such as text and speech, i.e., voice recognition. Unlike other neural networks, all inputs of RNN are connected, and all processed information will be memorized during the training process, which makes it very suitable for processing natural language.

Although neural network technology was developed decades ago, the lack of data, computational speed, and cost at that time prevented deep learning from being successfully brought into the business environment. However, with the significant increase in computing speed, reduction in computing cost, and maturity of the algorithms, technologies such as Deep Learning have been applied to business environments more frequently.

12.6.4.1 Convolutional Neural Networks

CNNs [21] are a type of deep learning algorithm widely used in computer vision tasks. They are particularly useful for image classification, object detection, and image segmentation. CNNs are inspired by the organization and functionality of the visual cortex in animals. The key idea behind CNNs is to automatically learn hierarchical representations of visual data. CNNs consist of multiple layers, including convolutional layers, pooling layers, and fully connected layers. Convolutional layers are the core building blocks of CNNs. They apply a set of learnable filters to the input image, which performs convolution operations. These filters capture different image features, such as edges, textures, and shapes. By applying multiple filters, the network can learn to recognize more complex patterns and objects. Pooling layers reduce the spatial dimensions of the input volume, which helps reduce computational complexity and control overfitting. The most common pooling operation is max pooling, which selects the maximum value within a neighborhood.

Fully connected layers are typically placed at the end of the network and are responsible for making the final predictions. These layers connect every neuron in one layer to every neuron in the next, enabling the network to learn complex relationships between features extracted by earlier layers. During the training process, CNNs are trained on labeled images, adjusting its parameters to minimize the difference between predicted outputs and true labels. The network is presented with labeled training images, and it adjusts its parameters to minimize the difference between the predicted outputs and the true labels. Figure 12.7 shows the workflow of a typical CNN.

One of the key advantages of CNNs is their ability to automatically learn feature representations from raw data, without the need for manual feature engineering. This makes CNNs highly effective at handling complex visual patterns and achieving state-of-the-art performance in various computer vision tasks. CNNs have revolutionized computer vision and are now essential for many AI applications involving image and video analysis.

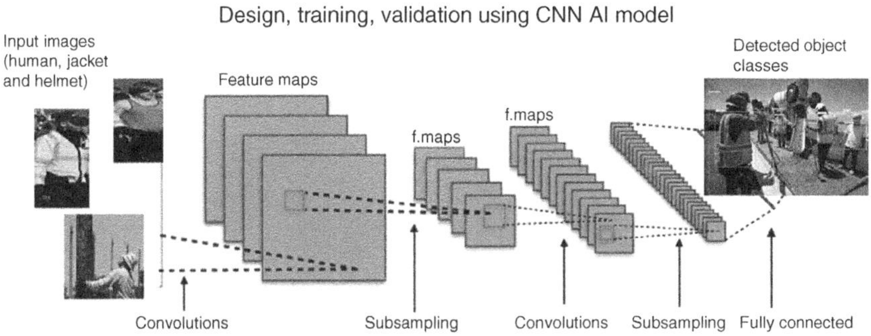

Figure 12.7 Convolutional Neural Network (CNN) [Adapted from Convolutional Neural Network: An Overview, https://www.analyticsvidhya.com/blog/2022/01/convolutional-neural-network-an-overview/, last accessed on October 09, 2024. Convolutional Neural Network: A standard convolutional neural network, By Aphex34 (Own work) [Alisdare Hickson/Flickr/CC BY-SA 2.0 Communications Office/Flickr/CC BY 2.0]

While CNNs are renowned for their prowess in image and object recognition, their applicability extends far beyond the visual realm. They can be effectively used in various domains, including NLP for text classification, translation, and text summarization, Time Series analysis for financial forecasting, and anomaly detection in sensor data or network traffic. As illustrated in the previous chapter, CNNs were used in conjunction with Finger Printing Technology for location tracking in the smart RFID system. CNNs are a versatile tool with the potential to revolutionize various fields beyond image processing. Their ability to learn complex patterns from structured data makes them well-suited for solving a wide range of problems.

12.6.4.2 Long Short-Term Memory

Similar to CNN, Long Short-Term Memory (LSTM) is popular and useful algorithm but for time series data. It is a powerful type of RNN specifically designed to handle sequential data, such as text, speech, or time series as explained earlier. Traditional RNNs struggle with capturing long-term dependencies in data, meaning they have difficulty remembering information from earlier parts of a sequence when processing later parts. This is due to the "vanishing gradient" problem, where signals used for learning become very small as they propagate through the network. LSTMs overcome this limitation by introducing "memory cells" that act like internal storage units, capable of holding information over extended periods. These memory cells are controlled by "gates," which act like filters, allowing the network to selectively remember or forget information based on its importance. The input gate controls which information from the current input is allowed to enter the memory cell, the forget gate controls which information from the

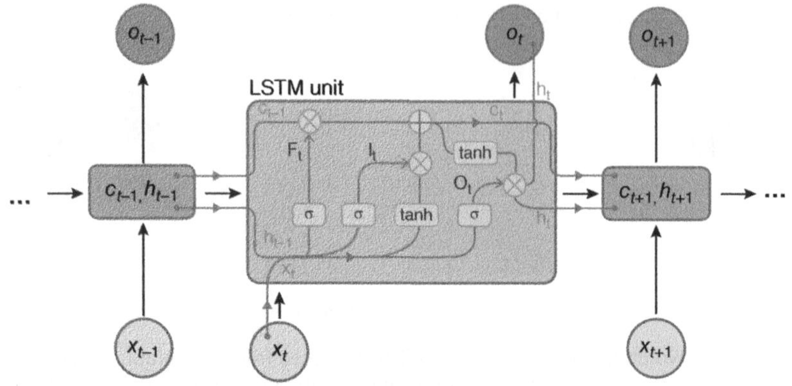

Figure 12.8 Standard Long Short-Term Memory (LSTM) neural network architecture for time series forecasting [fdeloche/Wikipedia/https://commons.wikimedia.org/wiki/File:Long_Short-Term_Memory.svg#/media/File:Long_Short-Term_Memory.svg, last accessed on October 09, 2024 /CC BY 4.0]

previous memory cell is forgotten, and the output gate controls which information from the memory cell is outputted to the next layer as shown in Figure 12.8. This unique combination of memory cells and gates allows LSTMs to excel at capturing long-range dependencies in sequential data, making them ideal for tasks like language translation, text summarization, sentiment analysis, speech recognition, and time series analysis. In essence, LSTMs are powerful tools for handling complex patterns and relationships in data where order and context matter, making them a crucial component of many modern AI applications.

12.6.5 Other Mainstream AI Algorithms

Apart from CNNs and LSTMs, there are a few strong contenders emerging, each with its own strengths and potential for widespread adoption.

12.6.5.1 Transformers

These powerful neural network architectures have already revolutionized NLP, powering large language models like GPT-3 and LaMDA, capable of generating human-quality text, translating languages, and answering questions. Researchers are now exploring their use in other domains, including computer vision, time series analysis, and even scientific discovery. Transformers are known for their ability to handle large amounts of data efficiently, making them suitable for training on massive datasets.

12.6.5.2 Explainable AI (XAI)

As AI systems become more complex, the need for explainability is growing. XAI aims to make AI models more transparent and understandable, allowing users to understand how decisions are made. This is crucial for building trust in AI systems, particularly in critical domains like healthcare, finance, and law. XAI also addresses ethical concerns related to bias and fairness in AI.

12.6.5.3 Federated Learning

This approach allows AI models to be trained on decentralized data, without requiring individual data to be shared with a central server. This is particularly relevant for edge computing, where AI models are deployed on devices like smartphones or IoT sensors. Federated learning helps address concerns about data privacy and security, making AI more accessible and trustworthy.

12.6.5.4 Continual Learning

AI systems need to be able to adapt to changing data and environments. Continual learning allows models to learn from new data without forgetting previously learned information. This is crucial for applications like autonomous vehicles, where AI systems need to learn from new experiences and adapt to changing road conditions.

12.6.5.5 Hybrid Architectures

Combining different architectures, like CNNs and LSTMs, or CNNs and transformers, is becoming increasingly common. This allows AI systems to leverage the strengths of each architecture for specific tasks. For example, a CNN might be used to extract features from an image, which are then fed into an LSTM or transformer to generate a caption. This approach allows for more tailored solutions that can handle complex tasks requiring diverse data types and processing capabilities. As an extension of the design example from the Smart Lamp Post (SLP) project, this IoT design is expanded to incorporate AI technology for predictive maintenance, using historical data to predict its current or future operating conditions. A hybrid AI approach, utilizing Cluster-LSTM with K-Means and LSTM algorithms, was employed for anomaly prediction and detection [22]. The K-Means is used to cluster the output features based on the historical data consisting of electrical and environmental parameters to set the norms (see Figure 12.9). Then, the LSTM is utilized to predict the current tor future value of the output parameters. Those output features are checked against the norms to determine the anomalies as shown in Figure 12.10. The accuracy of anomalies prediction and detection can be increased than by using a single approach.

The future of AI is likely to be characterized by a diverse range of architectures, each optimized for specific tasks and domains. Hybrid approaches, combining the strengths of different algorithms, will likely become more common. There will also

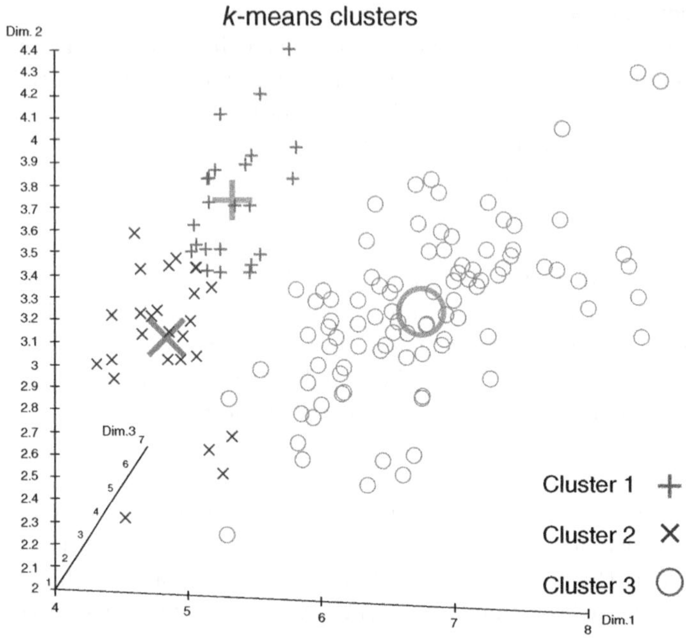

Figure 12.9 Clustering of input data [Chire/Wikipedia/https://commons.wikimedia.org/wiki/File:Iris_Flowers_Clustering_kMeans.svg#/media/File:Iris_Flowers_Clustering_kMeans.svg, last accessed on October 09, 2024/CC BY 4.0]

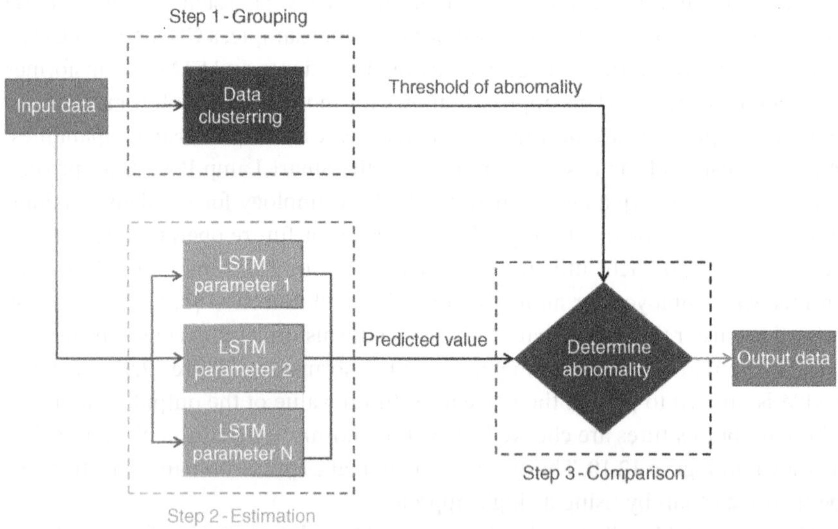

Figure 12.10 Design flow of Clustering-LSTM AI algorithm [courtesy of Lexiwave Technology (Hong Kong) Limited]

be a strong emphasis on explainability, trust, privacy, and the ability to adapt to changing data and environments. The AI landscape is constantly evolving, and the "mainstream" algorithms of tomorrow will be shaped by the challenges and opportunities that emerge in the years to come.

12.7 Edge Computing

Edge computing [23] represents a significant shift in the way data is process and managed. Instead of relying solely on centralized cloud servers, edge computing brings computation and data storage closer to the source of the data, right at the "edge" of the network. This decentralization approach offers several key advantages.

One of the most notable benefits of edge computing is the reduction in latency and improvement in responsiveness. By processing data locally, the time it takes for information to travel back and forth is significantly reduced. This makes edge computing ideal for real-time applications where quick responses are essential, such as autonomous vehicles or industrial automation. Furthermore, edge computing leads to efficient bandwidth usage. Since data is processed locally, there's less need to send it across vast distances, resulting in faster data transfer and reduced strain on network infrastructure. This is particularly important in environments with limited bandwidth or where data transfer costs are a concern.

Edge computing also enhances security by keeping data close to its origin, reducing the risk of security breaches during transmission. This is particularly important for sensitive information, where privacy is paramount. Another key advantage of edge computing is its ability to provide offline functionality. Edge devices can operate independently, even when Internet connectivity is unavailable. This is crucial for applications in remote or challenging environments where reliable Internet access is limited, such as disaster relief efforts or remote monitoring systems.

The applications of edge computing are vast and continue to expand across various industries. In the IoT, sensors and devices, such as the Tracking Watch introduced in Chapter 8, can process data in real time, locally and only send the final results to the server. In autonomous vehicles, real-time decision-making and navigation are made possible by processing data from sensors and cameras directly on the vehicle. Virtual and augmented reality experiences are enhanced by processing data locally, reducing latency and creating a more seamless interaction. Content delivery networks leverage edge computing to deliver content faster and more efficiently to users around the world. In healthcare, real-time data analysis can improve patient care, enabling faster diagnoses and more personalized treatment plans.

The integration of embedded AI [24, 25] into edge devices further amplifies the potential of edge computing. Embedded AI allows devices to analyze data in real time, enabling faster and more intelligent decision-making. This opens up a new world of possibilities. Imagine smart homes that can anticipate needs, industrial automation that optimizes efficiency, and autonomous vehicles that navigate more safely and efficiently, all powered by embedded AI working at the edge. In healthcare, embedded AI can analyze patient data in real time, enabling faster diagnoses, more personalized treatment plans, and improved overall patient care. Embedded AI can also enhance virtual/augmented reality experiences, creating more immersive and responsive interactions, and personalize content delivery in content delivery networks, ensuring users receive the most relevant information based on their individual preferences.

Immense potential is offered by edge computing, but challenges are also faced. Managing and securing a distributed network can be complex, and the deployment and scaling of edge infrastructure can be costly. However, the challenges are outweighed by the benefits of edge computing, and its adoption is expected to continue growing. As the technology matures, even more innovative applications and use cases are expected to emerge, transforming the way data is processed and managed, and creating a new era of intelligent and connected devices.

References

1 "What is a Distributed System?" last modified February 5, 2024, https://www.geeksforgeeks.org/what-is-a-distributed-system/

2 "Apache Kafka" last modified February 5, 2024, https://kafka.apache.org/

3 "Apache Hadoop" last modified February 5, 2024, https://cwiki.apache.org/confluence/display/hadoop

4 "MongoDB" last modified February 5, 2024, https://www.mongodb.com/

5 "Apache Flink" last modified February 5, 2024, https://flink.apache.org/

6 "Kubernetes" last modified February 5, 2024, https://kubernetes.io/

7 "Apache Mesos" last modified February 5, 2024, https://mesos.apache.org/

8 "What is a Cluster?" last modified February 5, 2024, https://www.techtarget.com/whatis/definition/cluster

9 "What is Azure?" last modified February 5, 2024, https://azure.microsoft.com/en-us/resources/cloud-computing-dictionary/what-is-azure/

10 Kusnetzky, D. (2011). *Virtualization: A Manager's Guide: Big Picture of the Who, What, and Where of Virtualization*. O'Reilly Media.

11 Erl, T. and Monroy, E. (2023). *Cloud Computing: Concepts, Technology, Security, and Architecture*. Pearson.

12 Bashir, I. (2023). *Mastering Blockchain: Inner Workings of Blockchain, from Cryptography and Decentralized Identities, to DeFi, NFTs, and Web3*, 4e. Packt Publishing.
13 "What is Distributed Ledger Technology (DLT)?" last modified February 5, 2024, https://www.techtarget.com/searchcio/definition/distributed-ledger
14 "23 Types of Consensus Mechanisms Used in Blockchains," last modified February 5, 2024, https://www.cryptolists.com/consensus-mechanisms/
15 Bishop, C. (2006). *Pattern Recognition and Machine Learning*, 195–222. Springer MIHE.
16 "What Is a Decision Tree?," last modified February 5, 2024, https://www.mastersindatascience.org/learning/machine-learning-algorithms/decision-tree/
17 "What is Clustering?," last modified February 5, 2024, https://developers.google.com/machine-learning/clustering/overview
18 "Bayesian Machine Learning," last modified February 5, 2024, https://www.datarobot.com/blog/bayesian-machine-learning/
19 "Artificial Neural Networks and its Applications," last modified February 5, 2024, https://www.geeksforgeeks.org/artificial-neural-networks-and-its-applications/
20 "Understanding K-Means Clustering in Machine Learning," last modified February 14, 2025, https://medium.com/towards-data-science/understanding-k-means-clustering-in-machine-learning-6a6e67336aa1
21 Goodfellow, I., Bengio, Y., and Courville, A. (2016). *Deep Learning: Chapter 9*. MIT Press Academic.
22 Zulfauzi, I.A. et al. (2023). "Anomaly detection using K-means and long-short term memory for predictive maintenance of large-scale solar (LSS) photovoltaic plant," *Energy Reports* 9: 154–158.
23 Lea, P. (2020). *IoT and Edge Computing for Architects: Implementing Edge and IoT Systems from Sensors to Clouds with Communication Systems, Analytics, and Security*, 2e. Packt Publishing.
24 Iodice, G.M. (2023). *TinyML Cookbook: Combine Artificial Intelligence and Ultra-low-power Embedded Devices to Make the World Smarter*, 2e. Packt Publishing.
25 Wang, X. et al. (2020). *Edge AI: Convergence of Edge Computing and Artificial Intelligence*, 1e. Springer.

Index

Introduction to Wireless System Design: From Circuits to Web-based Applications, First Edition.
Henry Lau, Ludy Liu, and Keith Chan.

d

Printed and bound by CPI Group (UK) Ltd, Croydon, CR0 4YY
17/09/2026
14964308-0002